W9-AUJ-147

UTOPIA IS CREEPY

ALSO BY NICHOLAS CARR

The Glass Cage

The Shallows

The Big Switch

Does IT Matter?

The Digital Enterprise (editor)

UTOPIA
IS
CREEPY

and Other Provocations

NICHOLAS CARR

W. W. NORTON & COMPANY

Independent Publishers Since 1923

New York   London

For information about permission to reproduce selections from this book,
write to Permissions, W. W. Norton & Company, Inc.,
500 Fifth Avenue, New York, NY 10110

For information about special discounts for bulk purchases, please contact
W. W. Norton Special Sales at specialsales@wwnorton.com or 800-233-4830

Manufacturing by Berryville Graphics
Book design by Lovedog Studio
Production manager: Anna Oler

ISBN 978-0-393-25454-9

W. W. Norton & Company, Inc.
500 Fifth Avenue, New York, N.Y. 10110
www.wwnorton.com

W. W. Norton & Company Ltd.
15 Carlisle Street, London W1D 3BS

1 2 3 4 5 6 7 8 9 0

To Nora and Henry

CONTENTS

Introduction: SILICON VALLEY DAYS   xv

UTOPIA IS CREEPY:
THE BEST OF ROUGH TYPE   1

THE AMORALITY OF WEB 2.0   3

MYSPACE'S VACANCY   10

THE SERENDIPITY MACHINE   12

CALIFORNIA KINGS   16

THE WIKIPEDIAN CRACKUP   18

EXCUSE ME WHILE I BLOG   21

THE METABOLIC THING   23

BIG TROUBLE IN SECOND LIFE   25

LOOK AT YOU!   28

DIGITAL SHARECROPPING   30

STEVE'S DEVICES   32

TWITTER DOT DASH   34

GHOSTS IN THE CODE   37

GO ASK ALICE'S AVATAR   39

LONG PLAYER   41

SHOULD THE NET FORGET?   47

THE MEANS OF CREATIVITY   49

VAMPIRES   50

BEHIND THE HEDGEROW, EATING GARBAGE   52

THE SOCIAL GRAFT   53

SEXBOT ACES TURING TEST   55

LOOKING INTO A SEE-THROUGH WORLD   56

GILLIGAN'S WEB   58

COMPLETE CONTROL   63

EVERYTHING THAT DIGITIZES MUST CONVERGE   66

RESURRECTION   69

ROCK-BY-NUMBER   71

RAISING THE VIRTUAL CHILD   73

THE IPAD LUDDITES   76

NOWNESS   79

CHARLIE BIT MY COGNITIVE SURPLUS   80

MAKING SHARING SAFE FOR CAPITALISTS   83

THE QUALITY OF ALLUSION IS NOT GOOGLE   86

SITUATIONAL OVERLOAD AND AMBIENT OVERLOAD   90

GRAND THEFT ATTENTION   93

MEMORY IS THE GRAVITY OF MIND   98

THE MEDIUM IS McLUHAN   102

FACEBOOK'S BUSINESS MODEL   107

UTOPIA IS CREEPY   108

SPINELESSNESS   110

FUTURE GOTHIC   112

THE HIERARCHY OF INNOVATION   116

RIP. MIX. BURN. READ.   121

LIVE FAST, DIE YOUNG, AND LEAVE
A BEAUTIFUL HOLOGRAM   126

ONLINE, OFFLINE, AND THE LINE BETWEEN   127

GOOGLE GLASS AND CLAUDE GLASS   131

BURNING DOWN THE SCHOOLHOUSE   133

THE ENNUI OF THE INTELLIGENT MACHINE   136

REFLECTIONS   138

WILL GUTENBERG LAUGH LAST?   140

THE SEARCHERS   144

ETERNAL SUNSHINE OF THE SPOTLESS AI   147

MAX LEVCHIN HAS PLANS FOR US   148

EVGENY'S LITTLE PROBLEM   151

THE SHORTEST CONVERSATION BETWEEN TWO POINTS   152

HOME AWAY FROM HOME   155

CHARCOAL, SHALE, COTTON, TANGERINE, SKY   160

SLUMMING WITH BUDDHA   162

THE QUANTIFIED SELF AT WORK   163

MY COMPUTER, MY DOPPELTWEETER   166

UNDERWEARABLES   168

THE BUS   170

THE MYTH OF THE ENDLESS LADDER   174

THE LOOM OF THE SELF   178

TECHNOLOGY BELOW AND BEYOND   179

OUTSOURCING DAD   181

TAKING MEASUREMENT'S MEASURE   182

SMARTPHONES ARE HOT   183

DESPERATE SCRAPBOOKERS   185

OUT OF CONTROL   187

OUR ALGORITHMS, OURSELVES   190

TWILIGHT OF THE IDYLLS   195

THE ILLUSION OF KNOWLEDGE   199

WIND-FUCKING   201

THE SECONDS ARE JUST PACKED   203

MUSIC IS THE UNIVERSAL LUBRICANT   207

TOWARD A UNIFIED THEORY OF LOVE   210

<3S AND MINDS   214

IN THE KINGDOM OF THE BORED,
THE ONE-ARMED BANDIT IS KING   216

THESES IN TWEETFORM   221

THE EUNUCH'S CHILDREN:
ESSAYS AND REVIEWS   227

FLAME AND FILAMENT   229

IS GOOGLE MAKING US STUPID?   231

SCREAMING FOR QUIET   243

THE DREAMS OF READERS   247

LIFE, LIBERTY, AND THE PURSUIT OF PRIVACY   255

HOOKED   260

MOTHER GOOGLE   264

THE LIBRARY OF UTOPIA   267

THE BOYS OF MOUNTAIN VIEW   279

THE EUNUCH'S CHILDREN   286

PAST-TENSE POP   292

THE LOVE THAT LAYS THE SWALE IN ROWS   296

THE SNAPCHAT CANDIDATE   314

WHY ROBOTS WILL ALWAYS NEED US   321

LOST IN THE CLOUD   325

THE DAEDALUS MISSION   329

Acknowledgments   343

Index   345

UTOPIA IS CREEPY

SILICON VALLEY DAYS

IT WAS A SCENE out of an Ambien nightmare: A jackal with the face of Mark Zuckerberg stood over a freshly killed zebra, gnawing at the animal's innards. But I was not asleep. The vision arrived midday, triggered by the Facebook founder's announcement—this was in the spring of 2011—that "the only meat I'm eating is from animals I've killed myself." Zuckerberg had begun his new "personal challenge," he told *Fortune* magazine, by boiling a lobster alive. Then he dispatched a chicken. Continuing up the food chain, he offed a pig and slit a goat's throat. On a hunting expedition, he put a bullet in a bison. He was "learning a lot," he said, "about sustainable living."

I managed to delete the image of the jackal-man from my memory. What I couldn't shake was a sense that in the young entrepreneur's latest pastime lay a metaphor awaiting explication. If only I could bring it into focus, piece its parts together, I might gain what I had long sought: a deeper understanding of the strange times in which we live. What did the predacious Zuckerberg represent? What meaning might the lobster's reddened claw hold? And what of that bison, surely the most symbolically resonant of American fauna? I was on to something. At the least, I figured, I'd be able to squeeze a decent blog post out of the story.

The post never got written, but many others did. I took up blogging early in 2005, just as it seemed everyone was talking about "the blogosphere." I discovered, after a little digging on GoDaddy, that the domain roughtype.com was still available (an uncharacteristic oversight by pornographers), so I called my blog Rough Type. The name seemed to fit the provisional, serve-it-raw quality of online writing at the time. Blogging has since been subsumed into journalism—it's lost its personality—but back then it did feel like something new in the world, a literary frontier. The collectivist claptrap about "conversational media" and "hive minds" that came to surround the blogosphere missed the point. Blogs were crankily personal productions. They were diaries written in public, running commentaries on whatever the writer happened to be reading or watching or thinking about at the moment. As one of the form's pioneers, Andrew Sullivan, put it, "You just say what the hell you want." The style suited the jitteriness of the web, that needy, oceanic churning. A blog was critical impressionism, or impressionistic criticism, and it had the immediacy of an argument in a bar. You hit the Publish button, and your post was out there on the World Wide Web, for everyone to see.

Or to ignore. Rough Type's early readership was trifling, which, in retrospect, was a blessing. I started blogging without knowing what the hell I wanted to say. I was a mumbler in a loud bazaar. Then, in the summer of 2005, Web 2.0 arrived. The commercial internet, comatose since the dot-com crash of 2000, was up on its feet, wide-eyed and hungry. Sites like MySpace, Flickr, LinkedIn, and the recently launched Facebook were pulling money back into Silicon Valley. Nerds were getting rich again. But the fledgling social networks, together with the rapidly inflating blogosphere and the endlessly discussed Wikipedia, seemed to herald something bigger than another gold rush. They were, if you could trust the hype, the vanguard of a democratic revolution in media and communication—a revolution that would change society forever. A new age was dawning, with a sunrise worthy of the Hudson River School.

Rough Type had its subject.

THE GREATEST of America's homegrown religions—greater than Jehovah's Witnesses, greater than the Church of Jesus Christ of Latter-Day Saints, greater even than Scientology—is the religion of technology. John Adolphus Etzler, a Pittsburgher, sounded the trumpet in his 1833 testament *The Paradise within the Reach of All Men.* By fulfilling its "mechanical purposes," he wrote, the United States would turn itself into a new Eden, a "state of superabundance" where "there will be a continual feast, parties of pleasures, novelties, delights and instructive occupations," not to mention "vegetables of infinite variety and appearance." Similar predictions proliferated throughout the nineteenth and twentieth centuries, and in their visions of "technological majesty," as the critic and historian Perry Miller wrote, we find the true American sublime. We may blow kisses to agrarians like Jefferson and tree-huggers like Thoreau, but we put our faith in Edison and Ford, Gates and Zuckerberg. It is the technologists who shall lead us.

Cyberspace, with its disembodied voices and ethereal avatars, seemed mystical from the start, its unearthly vastness a receptacle for America's spiritual yearnings and tropes. "What better way," wrote Cal State philosopher Michael Heim in 1991, "to emulate God's knowledge than to generate a virtual world constituted by bits of information?" In 1999, the year Google moved from a Menlo Park garage to a Palo Alto office, the Yale computer scientist David Gelernter wrote a manifesto predicting "the second coming of the computer," replete with gauzy images of "cyberbodies drift[ing] in the computational cosmos" and "beautifully-laid-out collections of information, like immaculate giant gardens." The millenarian rhetoric swelled with the arrival of Web 2.0. "Behold," proclaimed *Wired* in an August 2005 cover story: We are entering a "new world," powered not by God's grace but by the web's "electricity of participation." It would be a paradise of our own making, "manufactured by users." History's databases would be erased, humankind rebooted. "You and I are alive at this moment."

The revelation continues to this day, the technological paradise forever glittering on the horizon. Even money men have taken sidelines in starry-eyed futurism. In 2014, venture capitalist Marc Andreessen sent out a rhapsodic series of tweets—he called it a "tweetstorm"—announcing that computers and robots were about to liberate us all from "physical need constraints." Echoing John Adolphus Etzler (and also Karl Marx), he declared that "for the first time in history" humankind would be able to express its full and true nature: "We will be whoever we want to be. The main fields of human endeavor will be culture, arts, sciences, creativity, philosophy, experimentation, exploration, adventure." The only thing he left out was the vegetables.

Such prophesies might be dismissed as the prattle of overindulged rich guys, but for one thing: They've shaped public opinion. By spreading a utopian view of technology, a view that defines progress as essentially technological, they've encouraged people to switch off their critical faculties and give Silicon Valley entrepreneurs and financiers free rein in remaking culture to fit their commercial interests. If, after all, the technologists are creating a world of superabundance, a world without work or want, their interests must be indistinguishable from society's. To stand in their way, or even to question their motives and tactics, would be self-defeating. It would serve only to delay the wonderful inevitable.

The Silicon Valley line has been given an academic imprimatur by theorists from universities and think tanks. Intellectuals spanning the political spectrum, from Randian right to Marxian left, have portrayed the computer network as a technology of emancipation. The virtual world, they argue, provides an escape from repressive social, corporate, and governmental constraints; it frees people to exercise their volition and creativity unfettered, whether as entrepreneurs seeking riches in the marketplace or as volunteers engaged in "social production" outside the marketplace. "This new freedom," wrote law professor Yochai Benkler in his influential 2006 book *The Wealth of Networks*, "holds great practical promise: as a dimension of individual freedom; as a platform for better democratic participation; as a medium to foster a more critical and self-reflective culture; and, in an increasingly information-dependent global economy, as a mechanism to achieve improvements in human development everywhere." Calling it a revolution, he said, is no exaggeration.

Benkler and his cohorts had good intentions, but their assumptions were bad. They put too much stock in the early history of the web, when its commercial and social structures were inchoate, its users a skewed sample of the population. They failed to appreciate how the network would funnel the energies of the people into a centrally administered, tightly monitored information system organized to enrich a small group of businesses and their owners. The network would indeed generate a lot of wealth, but it would be wealth of the Adam Smith sort—and it would be concentrated in a few hands, not widely spread. The culture that emerged on the network, and that now extends deep into our lives and psyches, is characterized by frenetic production and consumption—smartphones have made media machines of us all—but little real empowerment and even less reflectiveness. It's a culture of distraction and dependency. That's not to deny the benefits of an efficient, universal system of information exchange. It is to deny the mythology that shrouds the system. And it is to deny the assumption that the system, in order to provide its benefits, had to take its present form.

Late in his life, the economist John Kenneth Galbraith coined the term "innocent fraud." He used it to describe a lie or a half-truth that, because it suits the needs or views of those in power, is presented as fact. After much repetition, the fiction becomes common wisdom. "It is innocent because most who employ it are without conscious guilt," Galbraith wrote. "It is fraud because it is quietly in the service of special interest." The idea of the computer network as an engine of liberation is an innocent fraud.

I LOVE a good gizmo. When, as a teenager, I sat down at a computer for the first time—a bulging, monochromatic terminal connected to a two-ton mainframe processor—I was wonderstruck. As soon as affordable PCs came along, I surrounded myself with beige boxes, floppy disks, and what used to be called "peripherals." A computer, I found, was a tool of many uses but also a puzzle of many mysteries. The more time you spent figuring out how it worked, learning its language and logic, probing its limits, the more possibilities it opened. Like the best of tools, it invited and rewarded curiosity. And it was fun, head crashes and fatal errors notwithstanding.

In the early nineties, I launched a browser for the first time and watched the gates of the web open. I was enthralled—so much territory, so few rules. But it didn't take long for the carpetbaggers to arrive. The territory began to be subdivided, strip-malled, and, as the monetary value of its data banks grew, strip-mined. My excitement remained, but it was tempered by wariness. I sensed that foreign agents were slipping into my computer through its connection to the web. What had been a tool under my own control was morphing into a medium under the control of others. The computer screen was becoming, as all mass media tend to become, an environment, a surrounding, an enclosure, at worst a cage. It seemed clear that those who controlled the omnipresent screen would, if given their way, control culture as well.

"Computing is not about computers any more," wrote MIT's Nicholas Negroponte in his 1995 bestseller *Being Digital*. "It is about living." By the turn of the century, Silicon Valley was selling more than gadgets and software. It was selling an ideology. The creed was set in the tradition of American techno-utopianism, but with a digital twist. The Valleyites were fierce materialists—what couldn't be measured had no meaning—yet they loathed materiality. In their view, the problems of the world, from inefficiency and inequality to morbidity and mortality, emanated from the world's physicality, from its embodiment in torpid, inflexible, decaying stuff. The panacea was virtuality—the reinvention and redemption of society in computer code. They would build us a new Eden not from atoms but from bits. All that is solid would melt into their network. We were expected to be grateful, and for the most part we were.

Our craving for regeneration through virtuality is the latest expression of what Susan Sontag, in *On Photography*, described as "the American impatience with reality, the taste for activities whose instrumentality is a machine." What we've always found hard to abide is that the world follows a script we didn't write. We look to technology not only to manipulate nature but to possess it, to package it as a product that can be consumed by pressing a light switch or a gas pedal or a shutter button. We yearn to reprogram existence, and with the computer we have the best means yet. We would like to see this project as heroic, as a rebellion against the tyranny of an alien power. But it's not that at all. It's a project born of anxiety. Behind it lies a dread that the messy, atomic world will rebel against us. What Silicon Valley sells and we buy is not transcendence but withdrawal. The screen provides a refuge, a mediated world that is more predictable, more tractable, and above all safer than the recalcitrant world of things. We flock to the virtual because the real demands too much of us.

"You and I are alive at this moment." That *Wired* story—its title was "We Are the Web"—nagged at me as the excitement over the rebirth of the internet intensified through the fall of 2005. The article was an irritant but also an inspiration. During the first weekend of October, I sat at my Power Mac G5 and hacked out a response. On Monday morning, I posted the result on Rough Type—a short essay with the portentous title "The Amorality of Web 2.0." To my surprise (and, I admit, delight), bloggers swarmed around the piece like phagocytes. Within days it had been viewed by thousands and had sprouted a tail of comments.

So began my argument with—what should I call it? There are so many choices: the digital age, the information age, the internet age, the computer age, the connected age, the Google age, the emoji age, the cloud age, the smartphone age, the data age, the Facebook age, the robot age, the posthuman age. The more names we pin on it, the more vaporous it seems. If nothing else, it is an age tailored to the talents of the brand manager. I'll just call it Now. It was through my argument with Now, an argument recorded in these pages, that I arrived at my own revelation, if only a modest, terrestrial one. What I want from technology is not a new world. What I want from technology are tools for exploring and enjoying the world that is—the world that comes to us thick with "things counter, original, spare, strange," as Gerard Manley Hopkins long ago described it.

My WordPress blogging program tells me I have published 1,608 posts at Rough Type. From that pile, I have extracted seventy-nine of my favorites for *Utopia Is Creepy*, beginning with 2005's "The Amorality of Web 2.0" and ending with "In the Kingdom of the Bored, the One-Armed Bandit Is King" from 2015. Complementing the posts are sixteen essays and reviews I wrote over roughly the same period. Also in the mix are fifty aphorisms, tweet-sized thoughts and fancies. Although most of these pieces were written to stand on their own, in combination they offer a chronicle of the last ten years that diverges in what I hope are stimulating ways from the prevailing account. We may all live in Silicon Valley now, but we can still act and think as exiles. We can still aspire to be what Seamus Heaney, in his poem "Exposure," called inner émigrés.

A dead bison. A billionaire with a gun. I guess the symbolism was pretty obvious all along.

Perfection does not exist;
to comprehend it is the triumph of human intelligence;
to desire to possess it, the most dangerous of follies.

—Alfred de Musset,
La Confession d'un Enfant du Siècle

The most unfree souls go west, and shout of freedom.

—D. H. Lawrence,
Studies in Classic American Literature

UTOPIA
IS CREEPY

The Best of Rough Type

THE AMORALITY OF WEB 2.0

October 3, 2005

THE WORLD WIDE WEB has always been a vessel of quasi-religious longing. And why not? For those seeking to transcend the material world, the web presents a readymade Promised Land. On the internet, we're all bodiless, symbols speaking to symbols in symbols. The early texts of web metaphysics, many written by thinkers associated with or influenced by the post-sixties New Age movement in California, are rich with a sense of impending spiritual release. They describe the passage into the cyber world as a process of personal and communal unshackling, a journey that frees us from the chains of mind, community, and identity, allows us to transcend our meager bodies. We become free-floating netizens in an enlightened, almost seraphic realm.

But as the web matured during the late 1990s, the dreams of a digital awakening went unfulfilled. The net turned out to be more about commerce than consciousness, more mall than commune. And when the new millennium arrived, it brought not a new age but a dispiritingly commonplace popping of a bubble of earthly greed. Somewhere along the way, the money changers had taken over the temple. The internet had transformed many things, but it had not transformed us.

The New New Age

The yearning for a higher consciousness didn't burst with the bubble. Web 1.0 may have turned out to be spiritual vaporware, but now we have the hyper-hyped upgrade: Web 2.0. In a new profile of the influential technology publisher Tim O'Reilly, *Wired* writer Steven Levy suggests that "the idea of collective consciousness is becoming manifest in the internet." He quotes O'Reilly: "The internet today is so much an echo of what we were talking about at Esalen in the '70s—except we didn't know it would be technology-mediated." Levy then asks, rhetorically, "Could it be that the internet—or what O'Reilly calls Web 2.0—is really the successor to the human potential movement?"

Levy's article appears in the afterglow of Kevin Kelly's ecstatic "We Are the Web" in *Wired*'s August issue. A *Whole Earth Catalog* editor before he helped launch *Wired*, Kelly serves as a nexus between hippie and hacker, a human fiber-optic cable beaming Northern Californian utopianism between generations. In his new article, a cover story, he surveys the recent history of the internet, from the Netscape IPO ten years ago, and concludes that the net has become a "magic window" that provides a "spookily godlike" perspective on existence. "I doubt angels have a better view of humanity."

And it's only "the Beginning." Soon, says Kelly, the web will grant us not only the vision of gods but also their powers. It is becoming the operating system "for a megacomputer that encompasses the internet, all its services, all peripheral chips and affiliated devices from scanners to satellites, and the billions of human minds entangled in this global network. This gargantuan Machine already exists in a primitive form. In the coming decade, it will evolve into an integral extension not only of our senses and bodies but our minds." We are being reborn, or at least retooled. "There is only one time in the history of each planet when its inhabitants first wire up its innumerable parts to make one large Machine. Later that Machine may run faster, but there is only one time when it is born. You and I are alive at this moment."

This isn't the language of exposition. It's the language of rapture.

The Cult of the Amateur

I'm all for seeking transcendence, whether it's by going to Sunday Mass or sitting at the feet of the Maharishi or gazing into the pixels of a liquid-crystal display. One gathers one's manna where one can. And if there's a higher consciousness to be found, then by all means let's get elevated. My problem is this: When we view the web in religious terms, when we imbue it with our personal yearning for transcendence, we can no longer see it objectively. By necessity, we have to look at the net as a moral force, not as a collection of inanimate hardware and software cobbled together by imperfect human beings. No decent person wants to worship an amoral conglomeration of technology.

And so all the things that Web 2.0 represents—participation, collectivism, virtuality, amateurism—become unarguably good things, things to be nurtured and applauded, emblems of progress toward enlightenment. But is it really so? Is there a counterargument to be made? Might, on balance, the practical effect of Web 2.0 on society and culture be bad, not good? To see Web 2.0 as a moral force is to deafen oneself to such questions.

Let me bring the discussion down to a brass tack. If you read anything about Web 2.0, you'll inevitably find praise heaped upon Wikipedia as a glorious manifestation of "the age of participation." Wikipedia is an open-source encyclopedia; anyone who wants to contribute to its construction can add an entry or edit an existing one. Tim O'Reilly says that Wikipedia marks "a profound change in the dynamics of content creation"—a leap beyond the Web 1.0 model of Britannica Online. To Kevin Kelly, Wikipedia shows how the web is allowing us to pool our individual brains into a great collective mind. It's a harbinger of the Machine.

In theory, Wikipedia is a beautiful thing—it *has* to be a beautiful thing if the web is leading us to a higher consciousness. In reality, Wikipedia isn't very good at all. Certainly, it's useful. I consult it all the time to get a quick gloss on a subject. But at a factual level it's unreliable, and the writing is often appalling. Take, for instance, this section from Wikipedia's biography of Bill Gates, excerpted verbatim:

Gates married Melinda French on January 1, 1994. They have three children, Jennifer Katharine Gates (born April 26, 1996), Rory John Gates (born May 23, 1999) and Phoebe Adele Gates (born September 14, 2002).

In 1994, Gates acquired the Codex Leicester, a collection of writings by Leonardo da Vinci; as of 2003 it was on display at the Seattle Art Museum.

In 1997, Gates was the victim of a bizarre extortion plot by Chicago resident Adam Quinn Pletcher. Gates testified at the subsequent trial. Pletcher was convicted and sentenced in July 1998 to six years in prison. In February 1998 Gates was attacked by Noël Godin with a cream pie. In July 2005, he solicited the services of famed lawyer Hesham Foda.

According to *Forbes*, Gates contributed money to the 2004 presidential campaign of George W. Bush. According to the Center for Responsive Politics, Gates is cited as having contributed at least $33,335 to over 50 political campaigns during the 2004 election cycle.

Excuse me for stating the obvious, but this is garbage, a hodge-podge of dubious factoids that adds up to something far less than the sum of its parts.

And here's Wikipedia on Jane Fonda's life:

Her nickname as a youth—Lady Jane—was one she reportedly disliked. She traveled to Communist Russia in 1964 and was impressed by the people, who welcomed her warmly as Henry's daughter. In the mid-1960s she bought a farm outside of Paris, had it renovated and personally started a garden. She visited Andy Warhol's Factory in 1966. About her 1971 Oscar win, her father Henry said: "How in hell would you like to have been in this business as long as I and have one of your kids win an Oscar before you do?" Jane was on the cover of *Life* magazine, March 29, 1968.

While early she had grown both distant from and critical of her father for much of her young life, in 1980, she bought the play "On Golden Pond" for the purpose of acting alongside her father—hoping he might win the Oscar that had eluded him throughout his career. He won, and when she accepted the Oscar on his behalf, she said it was "the happiest night of my life." Director and first husband Roger Vadim once said about her: "Living with Jane was difficult in the beginning . . . she had so many, how do you say, 'bachelor habits.' Too much organization. Time is her enemy. She cannot relax. Always there is something to do." Vadim also said, "There is also in Jane a basic wish to carry things to the limit."

This is worse than bad, and it is, unfortunately, representative of the slipshod quality of much of Wikipedia. Remember, this emanation of collective intelligence is not just a couple of months old. It's been around for nearly five years and has been worked over by many thousands of diligent contributors. At this point, it seems fair to ask exactly when the intelligence in "collective intelligence" will begin to manifest itself. When will the great Wikipedia get good? Or is "good" an old-fashioned concept that doesn't apply to emergent phenomena like communal online encyclopedias?

The promoters of Web 2.0 venerate the amateur and distrust the professional. We see it in their unalloyed praise of Wikipedia, and we see it in their worship of open-source software and myriad other examples of collectivist creativity. Perhaps nowhere, though, is their love of amateurism so apparent as in their promotion of blogging as an alternative to what they sneeringly call "mainstream media." Here's O'Reilly: "While mainstream media may see individual blogs as competitors, what is really unnerving is that the competition is with the blogosphere as a whole. This is not just a competition between sites, but a competition between business models. The world of Web 2.0 is also the world of what Dan Gillmor calls 'we, the media,' a world in which 'the former audience,' not a few people in a back room, decides what's important."

I like blogs, and I like blogging. But I'm not blind to the limits and flaws of the blogosphere—its superficiality, its emphasis on opinion over reporting, its echolalia, its tendency to reinforce rather than challenge ideological polarization and extremism. All the same criticisms can (and should) be aimed at segments of the mainstream media. But at its best the mainstream media is able to do things that are different from—and, yes, more important than—what bloggers can do. Those despised "people in a back room" can fund in-depth reporting and research. They can underwrite journalistic investigations that can take months or years to reach fruition—or that may fail altogether. They can hire and pay talented people who would not be able to survive as sole proprietors on the internet. They can employ editors and proofreaders and fact checkers and other unsung protectors of quality work. They can place opposing ideologies on the same page. Forced to choose between reading blogs and subscribing to, say, the *New York Times*, the *Wall Street Journal*, and *The Atlantic*, I'll choose the latter. I'll take the professionals over the amateurs.

But I don't want to be forced to make that choice.

The Price of Free

Having pushed the word-count limits of a blog post, I at last come to my point. The internet is changing the economics of creative work—or, to put it more broadly, the economics of culture—and it's doing so in a way that may well restrict rather than expand our choices. Wikipedia may be a shadow of the Britannica in many ways, but because it's created by amateurs rather than professionals, it's free. And free trumps good all the time. So what happens to those poor saps who write and edit encyclopedias for a living? They go away. The same thing happens when blogs and other free online writing and photography go up against old-fashioned newspapers and magazines. Of course the mainstream media sees the blogosphere as a competitor. It *is* a competitor. And, given the economics of the competition, it may well turn out to be a superior competitor. The layoffs we've recently seen at major newspapers may be just the beginning, and those layoffs should be cause not for self-satisfied snickering but for mourning. Implicit in the ecstatic visions of Web 2.0 is the hegemony of the amateur. I can't imagine anything more frightening.

In "We Are the Web," Kelly writes that "because of the ease of creation and dissemination, online culture is *the culture.*" I hope he's wrong, but I fear he's right—or will come to be right.

Like it or not, Web 2.0, like Web 1.0 before it, is amoral. It's a set of technologies—a machine, not a Machine—that alters the forms and economics of production and consumption. It doesn't care whether its consequences are good or bad. It doesn't care whether it brings us to a higher consciousness or a diminished one. It doesn't care whether it brightens culture or dulls it. It doesn't care whether it leads us into a golden age or a dark one. So let's ditch the utopian rhetoric and see the thing as it is, not as we wish it would be.

MYSPACE'S VACANCY

March 20, 2006

WHEN AN ADULT PUTS his ear to the door of youth culture, he inevitably mistakes the noise for the signal—and usually misses the signal altogether. So we have media blogger Scott Karp reeling back in horror after a visit to the hip social network MySpace. It is, he tells us, "a DEEPLY DISTURBING place," rife with "sexually suggestive or explicit content." There's a hint of criminality in the air. It is "humanity in the raw."

Excuse me while I go sign up for an account.

What's fascinating about Karp's post is not his reaction to MySpace but his reaction to his reaction to MySpace. Having offered a moral critique—a visceral one—he goes wobbly. "I'm not going to do a moral critique of MySpace or Web 2.0 or anything else—that's not my gig," he says. Then he says it again, with caps: "let me repeat—this is NOT a moral critique. It's a practical, business critique." A wise retreat, I suppose. Moral critiques are uncool. They're the surest way to lose your web cred.

Still, I liked the outburst, the act of recoiling. It was real. The practical, business critique seems forced in comparison: "'Social media' may be all the rage, but 'society' functions best somewhere in between anarchy and fascism. Let it drift too far to one extreme, and things can get ugly. And when things get ugly, it's hard to sell advertising." That's automatic writing, and when it's not platitudinous it's wrong. Ugly is edgy, and the edge is where advertisers want to be. Did Paris Hilton lose her endorsement deals when her naughty video leaked onto the web? Of course not. She got bigger and better ones.

A lot of bloggers hammered Karp for being an alarmist, for questioning the social media orthodoxy. One went so far as to compare

MySpace to a bicycle: Kids can get hurt on both, right? So what's the big deal? Maybe I'm misremembering, but I think my old banana bike was a pretty wholesome toy, even with the mile-high wheelie bar. Riding it around the neighborhood with my friends was a way to get some exercise and fresh air, to see things in three dimensions, to escape "my space." MySpace seems a little different.

Fred Wilson, a blogging venture capitalist, sees in MySpace the signs of a great emancipation:

> We are at the dawn of the age of personalized media. The web has given the world a place where the audience is the publisher and what we are witnessing (and hopefully participating in) is the personalization of media. It will manifest itself in many strange and wonderful ways. And I am embracing it; for me, for my kids, and for the rest of my life.

I guess you see what you want to see. When I look around MySpace I don't see much that's strange and wonderful—or deeply disturbing, either. I wish I did. What I see is a dreary sameness, a vast assemblage of interchangeable parts. Everything feels secondhand: the pimps-and-hoes poses before the cameraphone, the cliché-choked babbling. It's sad to see so much effort put into self-expression with so little to express. Humanity in the raw? No, this is humanity boiled to blandness in the tin pot of personalization.

There was another blogger who responded to Scott Karp's post by comparing the effect of MySpace to that of Elvis's gyrating hips back in the fifties. The old folks didn't get it then, and they don't get it now. But MySpace isn't anything like Elvis. It's more like an Elvis impersonator lip-syncing "Love Me Tender" in a white jumpsuit on the Vegas Strip.

I'll tell you what scares me about MySpace: It's not how dangerous it is, but how safe.

THE SERENDIPITY MACHINE

May 18, 2006

WE ARE GIVEN THE words we need when we need them. "Serendipity" slipped into the language 250 years ago, in 1754, when Horace
Walpole, the novelist, coined the word in a letter he sent to an acquain-
tance, the diplomat Horace Mann. Walpole was inspired by a Per-
sian fairy tale called "The Three Princes of Serendip," about a group
of royal travelers who "were always making discoveries," in Walpole's
words, "of things which they were not in quest of."

It took a while for the word to become commonplace. Robert Merton and Elinor Barber, in their book *The Travels and Adventures of Serendipity*, report that they could find only 135 instances of "serendipity"
in print in the 200 years after Walpole coined it. Then, beginning in
the late 1950s, its use exploded. Between 1958 and 2000, it appeared
in the titles of 57 books, according to historian and word-watcher
Richard Boyle, and in the 1990s alone it popped up in newspapers
some 13,000 times. In a 2000 U.K. poll, "serendipity" topped a list
of the public's favorite words. "The English-speaking world has gone
overboard for the word," writes Boyle.

Apparently something happened a half-century ago that gave us a
keener appreciation of—or need for—the serendipitous. Maybe it had
something to do with the prosperous calm we were granted after two
calamitous world wars. Or maybe it was an offshoot of the let-the-sun-
shine-in sensibility of the beats and their hippie offspring.

No sooner did the word become popular than its debasement began.
It drifted quickly toward nebulousness and then, as Merton argued in
an afterword to his and Barber's book, vacuity:

For many, it appears, the very sound of serendipity rather more than its metaphorical etymology takes hold so that at the extreme it is taken to mean little more than a Disney-like expression of pleasure, good feeling, joy, or happiness. For those who have consulted dictionaries for the word, its typical appearance between serenade and serene may bring a sense of tranquility and unruffled repose. In any case, no longer a niche-word filling a semantic gap, the vogue word became a vague word.

Boyle brings the story up to date:

So it is that in 1992 the word *serendipity* was emblazoned on the cover of a catalogue for women's underwear without further explanation. That in 1999 a review of the autobiography of Sir Alec Guinness drew attention to the actor's "serendipitous writing style (sly, witty, elegant)." That in 2001 the following was to be seen on the Internet: "Serendipity: When love feels like magic you call it destiny. When destiny has a sense of humour you call it serendipity."

Then, during a speech a few weeks ago, Google CEO Eric Schmidt revealed his aspiration to create a service called Google Serendipity, which "tells you what to type." Soon after hearing Schmidt's talk, I serendipitously came across a blog post on the subject of—you guessed it—serendipity written by Steven Johnson, author of the book *Everything Bad Is Good for You*. (We are also given the books we need when we need them.) Johnson was annoyed by a sentimental little op-ed in the *St. Petersburg Times* by a journalism professor named William McKeen, who argued that serendipity has become an "endangered joy":

We have become such a directed people. We can target what we want, thanks to the Internet. Put a couple of key words into a search engine and you find—with an irritating hit or miss here and

there—exactly what you're looking for. It's efficient, but dull. . . .
It's all about time. So many inventions save us time—whether it's
looking for information, shopping for clothes or checking what's
on television. Time is saved, but quality is lost. When you know
what you want—or think you do—you lose the adventure of discovery, of finding something for yourself.

Hogwash, harrumphed Johnson:

I find these arguments completely infuriating. Do these people
actually use the web? I find vastly more weird, unplanned stuff
online than I ever did browsing the stacks as a grad student. . . .
Thanks to the connective nature of hypertext, and the blogosphere's
exploratory hunger for finding new stuff, the web is the greatest
serendipity engine in the history of culture. . . . It's no accident that
Boing Boing is the most popular blog online—it's popular because
it's an incredible randomizer, sending you off on all these crazy and
unpredictable paths.

No, Steven, said English professor Alan Jacobs, in a comment on
Johnson's post, McKeen's right and you're wrong:

My particular situation is that of a scholar, and I think Steven is—
what's the technical term?—nuts to think that I now have more
serendipity than I did before. When I used to rely on print dictionaries or encyclopedias, I would very often forget what I was looking for because, in thumbing the pages, I would stumble across
all sorts of interesting words or topics, which would lead me to
look up other interesting words or topics, along the way to which
I would be distracted by yet other words or topics that I had never
seen before.

Another commenter at Johnson's blog also begged to differ: "Boing
Boing's interesting-link bounty, diverse as it might be, cannot be called

'serendipity.' When you pointed your browser there you knew what you were gonna get: links to Wonderful Things. It's a carefully curated randomness."

I come down in the middle. I think Johnson is absolutely right, and utterly wrong. I do think the web has expanded the sum total of serendipity in the world. I come across a heck of a lot more random stuff today than I did before I went online. So, yes, the web may well be "the greatest serendipity engine in the history of culture." Still, I find myself agreeing with McKeen when he talks of the lack of "surprise" in internet surfing and, even more so, with the commenter when he talks about the web's "carefully curated randomness." Once you create an engine—a machine—to produce serendipity, you destroy the essence of serendipity. It becomes something expected rather than something unexpected. Looking for serendipity? Just follow these easy links! Serendipity becomes an end in itself rather than an unanticipated surprise that leads you, at best, to some greater insight or understanding. Serendipity should be a doorway, a portal to elsewhere, but on the web it's a kitschy picture stuck on your refrigerator with a magnet in the shape of a banana.

What we're seeing is the final stage in the demeaning of the word "serendipity," maybe of the very idea of the serendipitous. The last act will be the unveiling of Google Serendipity. That will mark the end of serendipity's brief travels and adventures, the moment when serendipity becomes a packaged good, delivered upon request, a few advertisements in tow.

CALIFORNIA KINGS

July 7, 2006

GOOGLE HAS A CURIOUS management arrangement. Chief executive Eric Schmidt shares power with founders Sergey Brin and Larry Page. The workings of the triumvirate have long been hidden behind Google's tightly woven corporate cloak, but a ray of light pierced the fabric today. The *Wall Street Journal* published a report on how the threesome collaborated in retrofitting a Boeing 767 that Brin and Page bought last year for their personal use.

The journal got the scoop from Leslie Jennings, a corporate jet designer who had been hired to remodel the plane for the Googlers:

> Mr. Jennings says that Messrs. Brin and Page "had some strange requests," including hammocks hung from the ceiling of the plane. At one point he witnessed a dispute between them over whether Mr. Brin should have a "California king" size bed, he says. Mr. Jennings says Mr. Schmidt stepped in to resolve that by saying, "Sergey, you can have whatever bed you want in your room; Larry, you can have whatever kind of bed you want in your bedroom. Let's move on." Mr. Jennings says Mr. Schmidt at another point told him, "It's a party airplane."

Google has reaped big PR benefits from stories in the press highlighting its cofounders' modest and frugal lives. Much has been made, for instance, of the fact that Brin and Page drive Toyota Prius hybrids. In a 2004 profile on the show *20/20*, Barbara Walters reported that "Larry Page and Sergey Brin are not your typical billionaires. In fact, if you type billionaire into Google, the picture that emerges—fancy cars, private jets, mansions, jewels, supermodel girlfriends—isn't anything you'd find in the lifestyle of the Google guys. Page drives a Prius, which costs around $21,000. Brin gets around for the most part on in-line skates, and he still lives in a rented apartment." That same year, the BBC wrote that "far from living an extravagant lifestyle, complete with yachts and private jets like fellow software leader Oracle boss Larry Ellison," Page and Brin "live modest, unassuming lifestyles. They don't even have sports cars, and instead are said to each drive a Toyota Prius, a plain-looking but rather environmentally friendly saloon." Last year, *Business Week* gushed, "Flash and ostentation cut no ice at Google. . . . The status vehicle of choice at the Googleplex is the Toyota Prius hybrid, which both co-founders Sergey Brin and Larry Page drive." *Playboy*, in the introduction to a 2004 interview with Brin and Page, wrote, "The two are unlikely billionaires. They seem uninterested in the accouterments of wealth. Both drive Priuses, Toyota's hybrid gas-and-electric car."

A Prius can go about 55 miles on a gallon of gas. A Boeing 767, by contrast, will burn about 7,500 gallons of jet fuel during a typical five-hour party flight. The size of the beds does not appear to have a measurable impact on fuel consumption.

THE WIKIPEDIAN CRACKUP

September 5, 2006

FULFILLING ITS DESTINY AS a commune, Wikipedia is splitting into factions. Lots of them. The Wikipedian sects include—and this is not a complete list—antistatusquoists, authorists, communalists, community-ists, darwikinists, encyclopedists, essentialists, eventualists, exclusionists, exopedianists, immediatists, incrementalists, metapedianists, politicists, rehabilists, statusquoists, sysopists, and wikipacifists. Many of these sects also have subsects. There are extreme statusquoists and moderate statusquoists as well as extreme antistatusquoists and moderate antista-tusquoists. Each group has its own ideology, its own set of beliefs about what an online encyclopedia should look like and how it should be run.

The greatest of the wiki-sects, by far, are the deletionists and the inclusionists. These two bands of epistemological warriors are locked in a fight that is going to determine the fate of Wikipedia. The adher-ents of inclusionism believe that there should be no constraints on the breadth of the encyclopedia, that Wikipedia should include any entry that any contributor wants to submit. An article on a small-town elementary school is no less worthy of inclusion than an article on Stanford University. An article on a little girl's pet turtle is as valid as one on class Reptilia. The supporters of deletionism believe in weeding out entries that they view as trivial or otherwise inappropriate for a serious encyclopedia. They don't want their grand production to be fouled with dross. Between the inclusionists and the deletionists is not a middle ground but a no man's land.

Here's how the encyclopedia itself describes the two camps:

Deletionism is a philosophy held by some Wikipedians that favors clear and relatively rigorous standards for accepting articles, templates or other pages to the encyclopedia. Wikipedians who broadly subscribe to this philosophy are more likely to request that an article that they believe does not meet such standards be removed, or deleted. Conversely, Wikipedians who believe that there ought to be a place for an article on almost any topic in Wikipedia, and that there should be few or no standards barring an article from it, are said to subscribe to inclusionism.

There is an Association of Inclusionist Wikipedians, with 207 members at the moment, and they march under the slogan *Wikipedia is not paper*. Because there are no physical constraints on a digital encyclopedia's size, they see no reason to limit the subjects covered. Let's focus on making each article as good as possible, they say, not on picking which articles should stay and which should be deleted. If someone wants to contribute an article about their butterfly collection, that's fine. When there's space for everything, nothing is unworthy of documentation. There is as well an Association of Deletionist Wikipedians, currently with 144 members. They have a slogan of their own: *Wikipedia is not a junkyard*. To them, Wikipedia needs to be seen as a whole, not just as an assortment of articles. Culling insignificant entries is, in the deletionist view, essential to improving the quality of the overall work. Only truly "notable" subjects deserve an encyclopedia's imprimatur. To publish trivia would degrade the entire enterprise.

To the inclusionists, Wikipedia is in essence a wiki. It's an example of an entirely new form for collecting knowledge, a form unbound by the practices of the past. To the deletionists, Wikipedia is in essence an encyclopedia. It's an example of an established form for collecting knowledge, with traditions that deserve respect. The ideological split is a manifestation of an identity crisis that has always been inherent in Wikipedia. From the start, the publication has pursued two conflicting goals: to be an open encyclopedia that anyone can edit, and to be a serious encyclopedia that is as good as the best print encyclopedia. In the early years of Wikipedia's existence, when it was viewed mainly as a curiosity, the tension between these goals was easy to overlook.

Nobody much cared. But as Wikipedia has become more popular—and as it has begun to be held to a higher standard of quality—the tension has reached the snapping point. The inclusionists' desire for openness and the deletionists' desire for seriousness may both be worthy goals. But, as the diametrically opposed philosophies of the two camps make clear, they are mutually exclusive goals. You can't be a deletionist and an inclusionist at the same time.

At a deeper level, the divide is yet another example of the fundamental epistemological crisis of our time: the battle between absolutists and relativists. The deletionists are absolutists. They believe that some subjects are simply more significant than others, that objective distinctions can and should be drawn among different kinds of knowledge. John Milton is more important than George Jetson. The inclusionists are relativists. No subject is inherently more significant than any other, they argue. It all depends on the viewer's perspective. John Milton will be more important than George Jetson for some people. But for others George Jetson will be the more worthy of study. There are no absolutes; all distinctions are subjective.

The conflict between the inclusionists and the deletionists is more than academic. Articles are being deleted, undeleted, and redeleted from Wikipedia all the time, and the criteria for what stays and what goes are sources of constant and often bitter debate. If the deletionist philosophy prevails, as I suspect it will, the inclusionist Wikipedia will be lost forever. All those articles about pet parakeets and high school football coaches and disused Canadian railway stations will be erased from the servers, tossed aside as unworthy of notice or preservation. We will never know what a Whitmanesque encyclopedia, an exuberant *Leaves of Grass* of facts and factoids encompassing everything, would look like. I lean toward absolutism, but in this case I'll make an exception. Let Wikipedia be Wikipedia.

EXCUSE ME WHILE I BLOG

October 16, 2006

BLOG. *BLOG.* SAY IT five times in a row, preferably out loud: Blog. Blog. Blog. Blog. *Blog.* Has there ever been an uglier word? You don't speak it so much as expectorate it. As if it carried some toxin that you had to get out of your mouth as quickly and forcefully as possible. Blog! I think it must have snuck past the dictionary guards in disguise. Clearly, it was meant to mean something very different. I'd guess it was intended to be a piece of low slang referring to some coarse bodily function.

Like: "Can we pull over at the next rest area? I really have to blog."

Or: "The baby was up all night blogging."

Or: "Oh, Christ, I think I just stepped in a blog."

Somehow it escaped its scatological destiny and managed to hitch itself, like a tick, to a literary pursuit. Who's to blame? According to Wikipedia, Peter Merholz is the culprit. A guy called Jorn Barger introduced the term "web log"—on December 17, 1997—but it was Merholz who "jokingly broke the word 'weblog' into the phrase 'we blog' in the sidebar of his blog Peterme.com in April or May of 1999. This was quickly adopted as both a noun and verb." A passing act of silliness for which we all must now suffer. Thank you, Peter Merholz.

It doesn't seem fair. No other form of writing is saddled with such a grisly name. No one feels ridiculous saying "I am a novelist" or "I am a reporter" or "I am an essayist." Hell, you can even say "I am an advertising copywriter," and it sounds fairly respectable. But "I am a blogger"? Even when you say it to yourself, you can hear the sniggers in the shadows.

Imagine that you, a blogger, have just become engaged to some lovely person, and you are now meeting that lovely person's lovely parents for the first time. You're sitting on the sofa in their living room, sipping a flute of bubbly.

"So," they ask, "what do you do?"

Shame engulfs you. You try to say "I am a blogger," but you can't. The word lodges in your throat, won't budge. Panicked, you take refuge in circumlocution: "Well, I kind of, like, write, um, little commentaries that I, like, publish on the internet."

"Little commentaries?"

"Yes, you know, commentaries."

"About what?"

"Well, generally, they're commentaries that comment on other commentaries."

"How fascinating."

You're getting deeper in the muck, but you can't stop yourself. "Yeah. Usually it starts with some news story, and then I and a whole bunch of other people, other commentarians, will start commenting on it, and it'll just go from there. I mean, imagine that there's this news story and that a whole bunch of mushrooms start sprouting from it. Well, I'm one of those mushrooms."

Face it: Even "fungus" is a nicer word than "blog." In fact, if I had the opportunity to rename blogs, I think I would call them fungs. Granted, it's not a model of mellifluousness either, but at least its auditory connotations tend more toward the sexual than the excretory. "I fung." "I am a funger." Such phrases would encounter no obstacle in passing my lips.

But "I am a blogger"? Sorry. Can't do it. It sounds too much like a confession. It sounds like something you'd say while sitting in a circle of strangers in a windowless, linoleum-floored room in the basement of a medical clinic or a Baptist church. And then you'd start sobbing, covering your face with your hands. And then the fat lady sitting beside you would put her hand on your back. "It's all right," she'd say. "We're all bloggers here."

THE METABOLIC THING

October 21, 2006

THE *WASHINGTON POST* HAS an expose today on the restrooms in Google's headquarters. "Every bathroom stall on the company campus holds a Japanese high-tech commode with a heated seat," the paper reports. "If a flush is not enough, a wireless button on the door activates a bidet and drying." Tacked up beside that button, apparently for entertainment purposes, is a sheet of paper with "a geek quiz that changes every few weeks and asks technical questions about testing programming code for bugs."

I'm reminded of what Danny Hillis, the distinguished computer scientist whose work on parallel processing paved the way for Google's computer system, once said about the human race:

> We're the metabolic thing, which is the monkey that walks around, and we're the intelligent thing, which is a set of ideas and culture. And those two things have coevolved together, because they helped each other. But they're fundamentally different things. What's valuable about us, what's good about humans, is the idea thing. It's not the animal thing.

A few years back, when Google's founders still felt free to express their true ambitions, Sergey Brin said to a reporter, "Certainly if you had all the world's information directly attached to your brain, or an artificial brain that was smarter than your brain, you'd be better off. Between that and today, there's plenty of space to cover." And, certainly, if you had an artificial brain that was smarter than your brain, you'd no longer need to be imprisoned in the monkey that walks around.

Those Japanese commodes are nice, but it's important to remember that they're transitional devices. We'll know that Google has truly fulfilled its vision when the Googleplex no longer needs toilets at all.

BIG TROUBLE IN SECOND LIFE

November 15, 2006

THE COMMERCIALIZATION OF SECOND LIFE has hit a speed
bump. A new software program, called CopyBot, allows residents of
the virtual world to make perfect copies of other residents' creations.
The knockoffs threaten the livelihoods of the many entrepreneurs who
have set up shop selling digital trinkets in the sprawling fairyland. As
one irate avatar, named Caliandras Pendragon, writes in the newsletter
Second Life Insider:

> Those people who are living the dream that is promoted in every
> article, of earning a RL [Real Life] income from SL [Second Life]
> creations, are now living a nightmare in which their source of
> income may soon be worthless. That's not to speak of big commer-
> cial companies who have paid anything up to 1,000,000 dollars
> to have their product reproduced in loving detail, who will dis-
> cover that every Tom, Dick or Harriet may rip off their creation
> for nothing—and then sell it as their own.

Even in an immaterial world, we remain materialists, covetous of our
possessions.

The furor took an ugly turn late last night when, according to the
Second Life Herald, a "seething mob" surrounded a CopyBot operation run by Second Life resident GeForce Go. The mob shouted that
Go was "ruining their Second Life." Fearing for her safety, Go closed
down her shop and sold her land. In a subsequent "tumultuous meeting with dozens of angry and fearful residents all talking at once,"

Second Life official Robin Linden "sought to allay fears of any further concern about mass copyright violations."

Now officially banned by Linden Lab, the company that operates Second Life, CopyBot was created, according to reporter Adam Reuters of Reuters' SL bureau, by a group called libsecondlife, "a Linden Lab-supported open source project." As the protests swelled, the group tried to remove the Copybot source code from the site. But it was too late. Copybot itself had been copied. The source code was for sale on the marketplace SLExchange, "raising the prospect," Reuters wrote, "that it could become widespread." The resident who is selling the bot, Prim Revolution, demonstrated the software's power by making a precise clone of Adam Reuters himself. Revolution defended the use of CopyBot, saying, "I think the idea of clones and bots is very cool, and I'll be adding more new features for things like automated go-go dancers at clubs."

Second Life watcher Kevin Lim, of the University of Buffalo's School of Informatics, compares CopyBot to the Replicator in *Star Trek*, which "can create any inanimate matter, as long as the desired molecular structure is on file." Lim notes that "after such a machine was invented, currency as we knew it ceased to function. Since everyone had the capability to create (replicate) anything they desire, capitalism as we knew it died, and the new dawn of perfect Marxian philosophy was adopted by the Federation."

CopyBot's subversive energies have been unleashed at the very moment that big companies are rushing in to Second Life to exploit its economic potential. *Business Week* reports that "savvy CEOs" are beginning to hang out in the virtual world, believing it "could emerge as the corporate environment of the future." IBM chief Sam Palmisano proudly declares, "I have my own avatar." In fact, the magazine notes, Palmisano has two Second Life avatars: "a casual Sam and a buttoned-down one." What's going to happen, one wonders, when a CopyBot-armed anarchist comes up to one of the Palmisano avatars and replicates it? It may be a dream of CEOs to clone themselves for posterity, but would they be happy about being cloned by someone else, an interloper who might inhabit their virtual identity and use it to spread mischief and confusion? I can't imagine a captain of industry being sanguine about seeing a copy of himself flying through the air wearing only a g-string or emerging from a virtual sex shop bearing a mammoth steely dan.

The CopyBot controversy seems to herald a deeper crisis in Second Life (and by extension the web at large)—a struggle over the identity, the very essence, of the virtual land. Will Second Life remain a freewheeling utopian community operating by its own rules, or will it become a commercial operation, a virtual shopping mall with lots of advertisements and staged PR events? Is CopyBot a ripoff machine that threatens to destroy Second Life, or is it a sharing machine that would save the community's soul?

LOOK AT YOU!

December 17, 2006

TIME'S ISSUE OF MAY 6, 1946, featured famed horse breeder and makeup magnate Elizabeth Arden on its cover. The day the magazine appeared on newsstands, twenty-two of Arden's thoroughbreds died in a stable fire in Chicago. The tragedy confirmed what everyone had suspected: To have your picture on the front page of the popular news-weekly was to be cursed. Whom the gods would destroy they first put on the cover of *Time.*

In a cruel trick this week, *Time* has saddled all of us with its cover jinx. The magazine announced its Man of the Year for 2006, and he is You. To illustrate its choice, *Time* put on its cover a picture of a computer screen with a mirrored mylar finish. Look at it and you see yourself.

To what do we owe this dubious honor? To Web 2.0, of course. Thanks to the internet, writes *Time*'s top editor, Rick Stengel, "the creators and consumers of user-generated content are transforming art and politics and commerce." They are "the engaged citizens of a new digital democracy." The mirror on the cover "literally reflects the idea that you, not we, are transforming the information age."

Web 2.0 is battering the Great Man theory of history, writes Lev Grossman in the cover story. Yes, great men were responsible for "many painful and disturbing things that happened in 2006"—from war to global warming to the PlayStation 3 shortage—but "look at 2006 through a different lens and you'll see another story, one that isn't about conflict or great men. It's a story about community and collaboration on a scale never seen before. . . . It's about the many wresting power from the few and helping one another for nothing and how that will not only change the world, but also change the way the world changes." Web 2.0 is "really a revolution."

Also part of the cover package is a fawning profile of YouTube founders Chad Hurley and Steve Chen. The duo, proclaims *Time*, are "the new demiurges of the online world." Demiurges? Someone might want to tell Lev Grossman that the Great Man theory is still alive and well in the pages of *Time*.

But it's the cover, really, that contains the subtlest thinking in the issue. Web 2.0, writes Grossman, provides "a chance for people to look at a computer screen and really, genuinely wonder who's out there looking back at them." The cover gives Grossman's words a wry twist, offering a much darker view of the radical personalization of culture. Peer into the cover's computer screen and all you see looking back at you is you. In a solipsistic world, every Lonely Girl is a Great Man.

DIGITAL SHARECROPPING

December 19, 2006

STRIP THE HAPPY-FACE EMOTICONS from the social web, and you're left with a sad-face truth: By putting the means of production into the hands of the masses but withholding from those same masses any ownership over the products of their work, the internet provides an incredibly efficient mechanism for harvesting the economic value of the free labor provided by the very many and concentrating it into the hands of the very few.

A new analysis of web traffic, published by the blog Read Write Web, underscores the point. Despite the explosion of web content over recent years, web traffic appears to be growing ever more concentrated. The top ten sites accounted for 40 percent of total page views in November 2006, up from 31 percent in November 2001. The greater concentration comes during a period when the number of domains on the web nearly doubled, from 2.9 million to 5.1 million.

The numbers don't tell the whole story, though. The centralization of traffic can largely be explained by the popularity of two social networking sites, MySpace and Facebook, which together accounted for 17 percent of all page views in November. The content of sites like MySpace and Facebook is created by the networks' members. It's the members who write the updates and messages, post the photographs, upload the videos and the songs. If we counted each member's "presence"—the sum total of his or her contributions—as a separate site, which in a sense it is, we would find no increase in the concentration of traffic. Each social network contains a diversity of content produced by millions of individuals.

What's really being concentrated, in other words, is not online content but the economic value of that content. MySpace, Facebook,

YouTube, TripAdvisor, and many other internet businesses have realized that they can give away the tools of production but maintain ownership over the resulting products. The members of social networks may think they're just "sharing" their thoughts, interests, and opinions with other members, but actually they're working for the companies that operate the networks. The millions of people who write restaurant reviews for Yelp or book reviews for Amazon don't get paid for their labor, but the content they create is a valuable asset for the companies.

It's a modern kind of sharecropping system. Like plantation owners in the American South after the Civil War, a social network gives each member a little plot of virtual land on which to cultivate an online presence, through the posting, for instance, of words and pictures, and then the social network collects the economic value of the member's labor through advertising (or, less frequently, through subscriptions or the sale of goods). The digital sharecroppers are generally happy, because their interest lies in self-expression or socializing, not in making money, and, besides, the economic value of each of their individual contributions is trivial. It's only by aggregating those contributions on a massive scale—on a web scale—that the business becomes lucrative. To put it a different way, the sharecroppers operate happily in an attention economy while their overseers operate happily in a cash economy. In this view, the attention economy of the digital realm does not operate separately from the cash economy of the real world. The attention economy is just a means of creating cheap inputs for the cash economy. Many share in the work, few in the profit.

STEVE'S DEVICES

January 10, 2007

IT'S HARD TO IMAGINE the pleasure Steve Jobs must get from sin-
glehandedly upstaging the entire Consumer Electronics Show. When
he introduced Apple's "revolutionary mobile phone" at Macworld in
San Francisco yesterday—the gadget is called (surprise!) iPhone—the
euphoric press coverage drowned out everything happening in Vegas.
There was just one moment during Jobs's slick two-hour presentation
when he went off script, but it was a telling one. His clicker failed, and
while he waited for his backstage minions to fix the glitch he launched
into a reminiscence about how, back in the day, he and Woz hacked
together a little device that could jam television signals. They took it
over to Berkeley and used it to mess with the minds of the privileged
college kids by interrupting their viewing of *Star Trek*.

Jobs hasn't changed at all. He's still jamming signals, and getting a
huge kick out of it.

It was fun to contrast Jobs's presentation with the one Bill Gates
gave at CES a day earlier. Thematically, Gates's talk was a replay of
his keynote at last year's CES. He's still pitching a "digital lifestyle"
that nobody wants. Last year, it involved having computer screens all
over your kitchen so you'd be able to track the movements of your
family members and watch a bunch of different video feeds simultaneously while sipping your morning joe. It was a vision of the homeowner as Captain Kirk manning the bridge. This year's was stranger
yet. Gates sang the praises of the Windows Home Server, suggesting
that people want nothing more than to be network administrators—
the homeowner not as Kirk but as Scotty. He then led the audience
into a mockup of the bedroom of the future, the walls of which were

covered entirely in computer screens. It seemed a little pervy, in an antiseptic way.

Gates wants to engineer systems. Jobs wants to make tools.

Jobs, in fact, couldn't possibly be more out of touch with today's Web 2.0 ethos, which is all about platforms, open systems, egalitarianism, and the erasing of the boundary between producer and consumer. Like the iPod before it, the iPhone is a little fortress ruled over by King Steve. It's as self-contained as a hammer. It's a happening staged for an elite of one. The rest of us are free to gain admission by purchasing a ticket for $500, but we're required to remain in our seats at all times while the show is in progress. User-generated content? Hah! We're not even allowed to change the freaking battery. In Jobs's world, users are users, creators are creators, and never the twain shall meet. Which is, of course, why the iPhone, like the iPod, is an exquisite device. Steve Jobs is not interested in amateur productions.

TWITTER DOT DASH

March 18, 2007

AND SO AT LAST, after passing through Email and Instant Messaging and Texting, we arrive in the land of Twitter. The birds are singing in the trees—they look like that robin at the end of *Blue Velvet*—and the air is so clean you can see yourself in it.

Twitter is the telegraph system of Web 2.0. Like Morse's machine, it limits messages to very brief strings of text. But whereas the telegraph imposed its limit through the market's will—priced by the word, telegraph messages got expensive fast—Twitter imposes its limit through the iron law of code. Each message may include no more than 140 characters. As you type your message—your "tweet," in Twitterese—in the Twitter messaging box, a counter lets you know how many characters you have left. (That last sentence wouldn't quite have made the cut. It has 141 characters. Faulkner and Proust would have been disasters as Twitterers.)

Only on the length of each message is a limit imposed. Because there's no charge to send a message and no restriction on the frequency of posting, you can send as many tweets as you want. The telegraph required you to stop and ask yourself: "Is this worth it?" Twitter says: "Everything's worth it!" You can also broadcast each tweet to as large an audience as you want. Twitter unbundles the blog, fragments the fragment. It turns texting into a mass medium.

And what exactly are we broadcasting? The minutiae of our lives and thoughts. The moment-by-moment answer to what is, in Twitterland, the most important question in the world: *What are you doing?* Or, to save a half-dozen characters: *What u doing?* Twitter is the medium of Narcissus. Not only are you the star of the show, but everything that happens to you, no matter how trifling, is a headline, a media event, a stop-the-presses bulletin. Quicksilver turns to amber.

Are you exhausted yet?

Dave Winer, the venerable blogger, has created a *New York Times* feed that pours the paper's headlines through the Twitter service, as if to prove that everything is equal in its 140-character frivolity. "All the news that's fit to twit," twitters Dave.

my dog just piddled on the rug! :-) [less than 10 seconds ago]

Seventeen killed in Baghdad suicide bombing [2 minutes ago]

OMG I cant believe I just ate 14 Double Stuf Oreos [3 minutes ago]

"Twitter is a representation of my stream of consciousness," writes one Twitterer. What used to happen in the privacy of the mind is now tossed into the public's bowl like so many Fritos. The broadcasting of the spectacle of the self has become a full-time job. Au revoir, Jean Baudrillard, your work here is done.

Like so many other Web 2.0 services, Twitter wraps itself and its users in an infantile language. We're not adults having conversations, or even people sending messages. We're tweeters twittering tweets. We're twitters tweetering twits. We're twits tweeting twitters. We're Tweety Birds.

I did! I did taw a puddy tat! [half a minute ago]

I tawt I taw a puddy tat! [1 minute ago]

Narcissism is the user interface for nihilism, and with artfully kitschy services like Twitter we're allowed both to indulge our self-absorption and to distance ourselves from it by acknowledging, with a coy digital wink, its essential emptiness. I love me! Just kidding!

The great paradox of social networking is that it uses narcissism as the glue for community. Being online means being alone, and being in an online community means being alone together. (Baudrillard: "the community has been liquidated and absorbed by communication.") The community is purely symbolic, a pixel-painted simulation conjured up by software to feed the modern self's bottomless hunger. Hunger for what? For verification of its existence? No, not even that. For verification that it has a role to play. Like other social media but more so, Twitter offers, to borrow a phrase from the philosopher John Gray, "a refuge from insignificance."

As I walk down the street with thin white cords hanging from my ears, as I look at the display of casuals in the window of the Gap, as I sit in a Starbucks sipping a chai latte served up by a pierced barista, I can't quite bring myself to believe I'm real. But if I send out to a theoretical audience of my peers 140 characters of text saying that I'm walking down the street, looking in a shop window, drinking tea, suddenly I become real. I have a voice. I play a part. I mean something, if only as an image in a fun house of images.

It's not, as somebody put it, "I Twitter, therefore I am." It's "I Twitter because I'm afraid I ain't."

As the physical world takes on more of the characteristics of a simulation, we seek reality in the simulated world. At least there we can be confident that the simulation is real. At least there we can be freed from the anxiety of not knowing where the edge between real and unreal lies. At least there we can find something to hold on to, even if it's nothing.

GHOSTS IN THE CODE

April 4, 2007

LAST MONTH, AMAZON.COM WAS granted a broad patent for computer systems that incorporate human beings into automated data processing—the type of cybernetic arrangement that underpins the company's Mechanical Turk crowdsourcing service. With Turk, a software programmer can write into a program a task that is difficult for a computer to do but easy for a person to carry out, such as identifying objects in photographs. At the point in the program when the "human input" is required, the task is posted to the Turk website where people carry it out for a small payment. The human input is then funneled back to the computer running the program. Mechanical Turk, in essence, turns people into code.

The patent, as Amazon describes it, covers "a hybrid machine/human computing arrangement which advantageously involves humans to assist a computer to solve particular tasks, allowing the computer to solve the tasks more efficiently." It specifies several applications of such a system, including speech recognition, text classification, image recognition, image comparison, speech comparison, transcription of speech, and comparison of music samples. Amazon also notes that "those skilled in the art will recognize that the invention is not limited to the embodiments described."

The patent goes into great detail about how the system might work in evaluating the skills and performance of the "human-operated nodes." The system might, for example, want to classify the human workers according to whether they are "college educated, at most high school educated, at most elementary school educated, [or] not formally educated." It also lays out an example of the system incorporating "multiple humans" to carry out a particular subtask, "each of the humans being identified as being capable of satisfying at least some of the associated criteria," and then synthesizing a result from the combined inputs.

The patent would appear to be a brilliant hedging strategy by Amazon. There may come a time when humans are so busy carrying out menial tasks for computer systems that they have neither the time nor the money to buy books and other goods online. If so, Amazon can still look forward to earning hefty fees as the employment agency for the posthuman economy.

GO ASK ALICE'S AVATAR

April 22, 2007

RED LIGHT CENTER, an online world where avatars disrobe and mount each other in ways we humans can only dream about, expanded the bounds of virtual vice on Friday by introducing digital dope. Members of the community can now, reports *Technology Review,* "enter a virtual rave and take virtual ecstasy, smoke a virtual joint, and even munch on some virtual mushrooms."

Far out.

According to Brian Shuster, chief executive of Utherverse, the company that runs Red Light Center, the site is introducing simulated recreational drugs as a public service. Getting wasted virtually, he suggests, will decrease people's desire to get wasted in real life:

> In a virtual environment, [peer] pressure shifts from trying actual drugs to experimenting with virtual drugs. Thus, users have a safe platform to explore the social aspects of drug use, without having to risk doing the actual drugs. By separating the social pressure from the real-world application, users have a totally revolutionary mechanism to deal with peer pressure, and actually to give in to peer pressure, without the negative consequences. Moreover, users of virtual drugs have reported the effects of these virtual drugs to be surprisingly realistic and lifelike. To the extent that users can enjoy both the social benefits of virtual drugs as well as the entertainment associated with drug use, all with no actual drug consumption, the value of taking actual drugs is diminished.

I bet if you were really, really high that might actually make sense.

There is a certain symmetry to the idea of virtual drugs. When the concept of virtual reality first arose at the end of the sixties, it was tightly connected to the drug culture. The consciousness-expanding hallucinations that might be conjured up by computers weren't so different from those that emanated from a tab of acid (or so it seemed at the time). Now that it's possible to get stoned in cyberspace, we've come full circle. Think about it: When avatars hallucinate, they must see the real world.

Whoa. I'm freaking myself out.

LONG PLAYER

May 20, 2007

I STARTED READING DAVID WEINBERGER's new book, *Every-thing Is Miscellaneous: The Power of the New Digital Disorder*, this weekend. But I didn't get far. In fact, I only reached the bottom of page nine, at which point I crashed into this passage about music:

> For decades we've been buying albums. We thought it was for artis-tic reasons, but it was really because the economics of the physi-cal world required it: Bundling songs into long-playing albums lowered the production, marketing, and distribution costs because there were fewer records to make, ship, shelve, categorize, alpha-betize, and inventory. As soon as music went digital, we learned that the natural unit of music is the track. Thus was iTunes born, a miscellaneous pile of 3.5 million songs from a thousand record labels. Anyone can offer music there without first having to get the permission of a record executive.

The natural unit of music is the track? Well, roll over, Beethoven, and tell Tchaikovsky the news.

There's a lot going on in that short passage, and most of it is wrong. Weinberger, an internet researcher at Harvard, does do a good job, though, of condensing into a few sentences what might be called the liberation mythology of the internet. This mythology is founded on a sweeping historical revisionism that conjures up an imaginary predigital world—a world of profound physical and economic con-straints—from which the web is now freeing us. We were enslaved, and now we are saved. In a bizarrely fanciful twist, the digital world is presented as a "natural" counterpoint to the supposed artificiality of the physical world.

I cast the book aside and fell to pondering. Actually what I did was sweep the junk off the dust cover of my sadly neglected turntable and pull out an example of one of those old, maligned "long-playing albums" from my shrunken collection of cardboard-sheathed LPs (arrayed alphabetically, by artist, on a shelf in a cabinet). I chose *Exile on Main St.* More particularly, I chose the unnatural bundle of tracks to be found on side three of *Exile on Main St.* Carefully holding the thin black slab of scratched, slightly warped, but still serviceable vinyl by its edges—you won't, I trust, begrudge me a pang of nostalgia for the thinginess of the predigital world—I eased it onto the spindle and set the platter to spinning.

If you're not familiar with *Exile on Main St.*, or if you know it only in a debauched digital form—whether as a single-sided plastic CD or a pile of miscellaneous undersampled iTunes tracks—let me explain that side three is the strangest yet most crucial of the four sides of the Rolling Stones' 1972 double-record masterpiece. The side begins, literally, in happiness—or Happyness—and ends, figuratively, in a dark night of the soul. (I realize that, today, it may be hard to imagine Mick Jagger having a dark night of the soul, but at the dawn of the gruesome seventies, with the wounds of Brian Jones's death, Marianne Faithfull's overdose, and Altamont's hippie apocalypse still fresh in his psyche, Mick was, I imagine, suffering an existential pain that neither a needle and a spoon nor even another girl could take away.)

But it's the middle tracks of the record that seem most pertinent to me in thinking about Weinberger's argument. Between Keith Richards's ecstatic, grinning-at-death "Happy" and Jagger's desperate, shut-out-the-lights "Let It Loose" come three offhand, wasted-in-the-basement songs—"Turd on the Run," "Ventilator Blues," and "I Just Want to See His Face"—that sound, in isolation, like throwaways. If you unbundled *Exile* and tossed these tracks onto the miscellaneous iTunes pile, they'd vanish, probably without a trace. I mean, who's going to buy "Turd on the Run" as a standalone track? And yet, in the context of the album that is *Exile on Main St.*, the three songs achieve a remarkable, tortured eloquence. They become *necessary.* They transcend their identity as tracks, and they become part of something larger. They become art.

Listening to *Exile*, or to any number of other long-playing bundles—*Revolver, Astral Weeks, Forever Changes, Blood on the Tracks, London Calling, Murmur, Tim, Bee Thousand* (the list, thankfully, goes on and on)—I could almost convince myself that the twenty-minute-or-so side of an LP is not just some ungainly by-product of the economics of the physical world but rather the "natural unit of music." As natural a unit, anyway, as the individual track.

The long-playing vinyl phonograph record, twelve inches in diameter and spinning at a lazy 33⅓ rpm, is, even today, a fairly recent technological development. (Recorded music in general is a fairly recent technological development.) After a few failed attempts to produce a long player in the early 1930s, the modern LP was introduced in 1948 by a record executive named Edward Wallerstein, then the president of Columbia Records, a division of William Paley's giant Columbia Broadcasting System. At the time, the dominant phonograph record had for about a half-century been the 78—a fragile, ten-inch shellac disk that could hold only about three or four minutes of music on a side.

Wallerstein, being a fiendish record executive, invented the long player as a way to bundle a lot of tracks onto a single disk in order to enhance the economics of the business and force customers to buy a bunch of crappy songs that they didn't want so they could get a quality track or two that they did want. Right? Wrong. Wallerstein in fact invented the long player because he wanted a format that would do justice to performances of classical works, which, needless to say, didn't lend themselves all that well to three-minute snippets on 78s.

Before his death in 1970, Wallerstein recalled how he pushed a team of talented Columbia engineers to develop the modern record album (as well as a device for playing it). In their first attempt, the engineers produced a record that could hold eight minutes of music on a side.

Wallerstein was unimpressed: "I immediately said, 'Well, that's not a long-playing record.'" He determined, by analyzing many classical compositions, that a long-playing record had to be able to contain, at minimum, seventeen minutes of music on a side. He calculated that, with two seventeen-minute sides, the LP format would be able to accommodate about 90 percent of all classical works. After another year or two of hard work, the engineers met Wallerstein's goal.

The long player was not, in other words, a commercial contrivance aimed at bundling together popular songs to the advantage of record companies and the disadvantage of consumers. It was a format painstakingly designed to provide people with a much better way to listen to recordings of classical works. In focusing on perfecting a medium for classical performances, Columbia actually sacrificed much of the lucrative pop market to its rival RCA, which at the time was developing a competing record format, the seven-inch, 45 rpm single.

A brief standards war ensued between the LP and the 45—it was called "the battle of speeds"—which ended with a happy technological compromise that allowed both formats to flourish. Record players were designed to accommodate both 33⅓ rpm albums and 45 rpm singles (and, for a while, the old 78s as well). The 45 format allowed consumers to buy individual hits for a relatively low price, while the LP provided them with the option of buying longer works for a somewhat higher price. Popular music soon followed classical in moving onto LPs, as musicians and record companies sought to maximize their sales and provide fans with more songs by their favorite artists. The introduction of the pop LP did not force customers to buy more songs than they wanted—they could still cherry-pick individual tracks by buying 45s. The LP *expanded* people's choices, giving them more of the music they clamored for.

Indeed, in suggesting that the long player resulted in a big pile of "natural" tracks being bundled together into artificial albums, Weinberger gets it precisely backward. It was the arrival of the LP that set off the explosion in the number of popular music tracks available to buyers. It also set off a burst of incredible creativity in popular music. Songwriters and bands began to take advantage of the new, extended format, turning the longer medium to their own artistic purposes. The result was a great flowering not only of wonderful singles, sold as 45s, but of carefully constructed sets of songs, sold as LPs. Was there also a lot of filler? Of course there was. When hasn't there been?

Weinberger also gets it backward in suggesting that the LP was a record industry ploy to constrain the supply of product—in order to have "fewer records to make, ship, shelve, categorize, alphabetize, and inventory." The album format, combined with the single format, brought a huge *increase* in the number of records—and, in turn, in the outlets that sold them. It unleashed a flood of recorded music. It's worth remembering that the major competitor to the record during this time was radio, which of course provided music for free. The best way—the only way—for record companies to compete against radio was to increase the number of records they produced, to give customers far more choices than any radio station could send over the airwaves. The long-playing album, in sum, not only gave buyers many more products to choose from, it gave artists many more options for expressing themselves. Everyone won. Far from being a constraint on the market, the physical format of the long player was a great spur to consumer choice and, even more important, to creativity. Who would unbundle *Exile on Main St.* or *Blonde on Blonde* or *Tonight's the Night*? Only a fool.

And yet it is the wholesale unbundling of LPs into a "miscellaneous pile" of digital song files that Weinberger would have us welcome as some kind of deliverance from decades of servitude to the long-playing album. One doesn't have to be an apologist for record executives—who in recent years have done a great job in demonstrating their cynicism and stupidity—to recognize that Weinberger is warping history in an attempt to prove an ideological point. Will the new stress on discrete digital tracks bring a new flowering of creativity in music? I don't know. Maybe we'll get a pile of gems, or maybe we'll get a pile of dreck. Probably we'll get a mix. But I do know that the development of the vinyl long-playing album, together with the vinyl single, was a development we should all be grateful for. We probably shouldn't rush out to dance on the album's grave.

As for the individual track being the "natural unit of music," that's a fantasy. Natural's not in it.

SHOULD THE NET FORGET?

August 26, 2007

THE *NEW YORK TIMES* RECENTLY GOT some search-engine-optimization religion, and now its articles, including old stories from the paper's vast archives, are much more likely to appear at or near the top of web searches. But the tactic has had an unintended consequence, writes the paper's public editor, Clark Hoyt:

> Long-buried information about people that is wrong, outdated or incomplete is getting unwelcome new life. People are coming forward at the rate of roughly one a day to complain that they are being embarrassed, are worried about losing or not getting jobs, or may be losing customers because of the sudden prominence of old news articles that contain errors or were never followed up.

Hoyt tells the story of one man, a former New York City official named Allen Kraus, who resigned his position back in 1991 after a run-in with a new boss. The resignation was briefly, and incorrectly, tied to a fraud investigation that was going on at the same time in his office. The *Times* published stories about the affair, including one with the headline "A Welfare Official Denies He Resigned Because of Inquiry"—and that headline now appears at the top of the results you get if you Google Kraus's name. Kraus is, with good reason, unhappy that his reputation is, sixteen years after the fact, again being tarnished.

Many other people find themselves in similar predicaments, and they are contacting the *Times* (and, one assumes, other papers and magazines) to ask that the offending stories be removed from the archives. The *Times* has routinely refused such requests, not wanting to get into the business of rewriting the historical record. Deleting the articles, one of the paper's editors told Hoyt, would be "like airbrushing Trotsky out of the Kremlin picture."

But if the *Times* is using search-engine-optimization techniques to push articles toward the top of search engine results, does it have any ethical obligation to ensure that old errors, distortions, or omissions in its reporting don't blight people's lives or reputations? *Times* editors are discussing the problem, writes Hoyt, and in some cases the paper has added corrections to old stories when proof of an error has been supplied.

The paper's predicament highlights a broader issue about the web's tenacious but malleable memory. Viktor Mayer-Schönberger, a public policy professor at Harvard, tells Hoyt that newspapers "should program their archives to 'forget' some information, just as humans do."

> Through the ages, humans have generally remembered the important stuff and forgotten the trivial, he said. The computer age has turned that upside down. Now, everything lasts forever, whether it is insignificant or important, ancient or recent, complete or overtaken by events. Following Mayer-Schönberger's logic, the *Times* could program some items, like news briefs, which generate a surprising number of the complaints, to expire, at least for wide public access, in a relatively short time. Articles of larger significance could be assigned longer lives, or last forever.

With search engine optimization—or SEO, as it's known—news organizations and other companies are actively manipulating the web's memory. They're programming the web to "remember" stuff that might otherwise have become obscure by becoming harder to find. So if we are programming the web to remember, should we also be programming it to forget—not by expunging information, but by encouraging certain information to drift, so to speak, to the back of the web's mind?

THE MEANS OF CREATIVITY

October 14, 2007

I WAS FLIPPING THROUGH the new issue of *The Atlantic* today when I came across this nugget from Ray Kurzweil: "The means of creativity have now been democratized. For example, anyone with an inexpensive high-definition video camera and a personal computer can create a high-quality, full-length motion picture." Yep. Just as the invention of the pencil made it possible for anyone to write a high-quality, full-length novel. And just as that saw in my garage makes it possible for me to build a high-quality, full-length chest of drawers.

VAMPIRES

October 23, 2007

> One should be sufficiently intelligent and interested to know a good deal about any person one comes into close contact with. About her. Or about him. But to try to know any living being is to try to suck the life out of that being. . . . It is the temptation of a vampire fiend, is this knowledge.
>
> —*D. H. Lawrence*

WHEN A MONSTER INVADES popular culture, the beast is almost always a projection of a communal fear, a manifestation of an unease that, deep-seated, has yet to emerge into the light of general consciousness. The most popular movie in America today is the vampire flick *30 Days of Night*. The books in Stephenie Meyer's vampire epic, *The Twilight Saga*, sell in the millions. One of the most-played Facebook games is Vampires, in which you earn points by sucking the blood of your friends. Vampires are everywhere.

The vampire myth is an ancient and potent one, going back at least four thousand years. "Although there are cultural variations in the various legends," writes one scholar, "there is always one defining trait of a vampire: a vampire sucks blood. It consumes another to sustain its own life." The vampire is unusual among monsters in using seduction rather than violence to get its way: "The vampire does not rip bodies apart or hack people to pieces, or stake them through the heart. He has to be invited in, and often has to persuade his victim to remove her cross before he makes a small, neat bite—a love bite, or the kiss of death—on her neck so that he can get the blood he needs to live." Victims give in willingly. Only after many visits do they begin to "experience a mysterious loss of selfness."

The strategy of today's internet giants might best be called vampiric. Their overriding goal is to *know* us, to transfer into their databases our informational life-blood. Their thirst is unquenchable. To survive, they must suck in ever more intimate details of our lives and desires. And we are not averse to the seduction. We embrace these companies, welcome them into our homes and our lives, because we desire the gifts they bear, the conveniences they provide. Yet even as we tilt our necks, we feel that mysterious loss of selfness. We sense that we're being emptied, that we're beginning to blur at the edges.

BEHIND THE HEDGEROW, EATING GARBAGE

October 28, 2007

WRITING IN *THE OBSERVER* today, privacy advocate Nick Rosen provides a list of practical recommendations for escaping electronic surveillance. When I say "practical," I'm kidding.

Here's how Rosen suggests purchasing a cell phone: "Travel to a town you have never visited before, to an area with no CCTV cameras, and ask a homeless person to buy a pay-as-you-go mobile phone for you. That way no shop will have your image on its CCTV. You will also have an anonymous mobile."

To avoid being tracked by utility companies, you should cancel your electric, gas, and water services and rely on "solar panels and rainwater harvesting." Rosen notes: "There are tens of thousands of people living without mains power, water or sewerage, in isolated cottages, behind hedgerows in caravans or in groups of yurts in country fields."

To get food, too, you need to "shop outside the system" if you want to avoid being monitored. You might barter for your groceries on the street, or join the "full-time scavengers living off food retrieved from supermarket bins."

As Rosen admits toward the end of his article, many of his suggestions will strike people as comical. I guess that's the point. We have allowed our lives to become open books for marketers and snoops, so much so that resistance at this point seems ridiculous. To go "off grid" now, you pretty much have to turn yourself into a counterespionage operative, a secret agent living in a yurt and nibbling the bruised leaves of a discarded cabbage.

THE SOCIAL GRAFT

November 6, 2007

"ONCE EVERY HUNDRED YEARS media changes," boy-coder turned big-thinker Mark Zuckerberg declared today at the Facebook Social Advertising Event in New York City. It's true. Look back over the last millennium or two, and you'll see that every century, like clockwork, there's been a big change in media. Cave painting lasted a hundred years, and then there was smoke signaling, which also lasted a hundred years, and of course there was the hundred years of yodeling, and then there was the printing press, which was invented almost precisely a hundred years ago, and so forth up to the present day—the day that Facebook picked up the hundred-year torch and ran with it. Quoth the Zuckster: "The next hundred years will be different for advertising, and it starts today."

Yes, today is the first day of the rest of advertising's life.

I like the way that Zuckerberg conflates "media" and "advertising." It cuts through the noise. It *simplifies*. Get over your mainstream-media hang-ups, granddads. Editorial is advertorial. The medium is the message from our sponsor.

Marketing is conversational, says Zuckerberg, and advertising is social. There is no intimacy that is not a branding opportunity, no friendship that can't be monetized, no kiss that doesn't carry an exchange of value. "Facebook's ad system," goes a company press release, "serves Social Ads that combine social actions from your friends—such as a purchase of a product or review of a restaurant—with an advertiser's message." What Zuckerberg calls the social graph is, it turns out, a platform for social graft.

The Fortune 500 is lining up for the new Facebook service. Coke's in, big time:

The Coca-Cola Company will feature its Sprite brand on a new Facebook Page and will invite users to add an application to their account called "Sprite Sips." People will be able to create, configure and interact with an animated Sprite Sips character. For consumers in the United States, the experience can be enhanced by entering a PIN code found under the cap of every 20 oz. bottle of Sprite to unlock special features and accessories. The Sprite Sips character provides a means for interacting with friends on Facebook. In addition, Sprite will create a new Facebook Page for Sprite Sips and will run a series of Social Ads that leverage Facebook's natural viral communications to spread the application across its user base.

Infect me. I'm yours.

It's a nifty system: First you get your customers to entrust their personal data to you, and then you not only sell that data to advertisers but you get the customers to be the vector for the ads. And what do the customers get in return? An animated Sprite Sips character to interact with.

SEXBOT ACES TURING TEST

December 8, 2007

RUSSIAN CROOKS HAVE UNLEASHED an artificial intelligence, called CyberLover, that poses as a would-be paramour in chat rooms and entices randy gentlemen to reveal personal information that can then be put to criminal use. Amazingly, the software appears to be successful in convincing targets that it's a real person—a sexpot rather than a sexbot. "The artificial intelligence of CyberLover's automated chats is good enough that victims have a tough time distinguishing the 'bot' from a real potential suitor," reports CNET, drawing on a study by security researchers. "The software can work quickly too, establishing up to ten relationships in thirty minutes."

Could it be that the Turing Test has finally been beaten—by a sex machine, no less—and that a true artificial intelligence is on the loose? Maybe so, but this breakthrough, like Barry Bonds's homer record, is going to have to carry an asterisk. Studies show that when people enter a state of sexual arousal their intelligence drops precipitously. I won't go so far as to say that CyberLover is cheating, but I will argue that it has an unfair advantage over other AI wannabes.

There's a bigger lesson in the sexbot's success. If you want to push artificial intelligence beyond human intelligence, you have two options: Make machines smarter or make people dumber. CyberLover suggests the latter path may prove the quickest route to the Singularity.

LOOKING INTO A SEE-THROUGH WORLD

January 31, 2008

THE CITIZENS OF BARROW GURNEY in southwestern England have asked that their village be erased from digital maps. Like many towns around the world, Barrow Gurney has been overrun by cars and trucks whose drivers robotically follow the instructions dispensed by GPS systems. The shortest route between two points sometimes runs right through once-quiet neighborhoods and formerly out-of-the-way hamlets.

A new generation of digital maps may make things worse. Connected directly to the internet, they provide drivers with a stream of real-time information about traffic congestion, accidents, and road construction. The debut of one of the new systems, called Dash Express, at this month's Consumer Electronics Show in Las Vegas led to claims that the new technology might "spell the end of traffic jams forever."

That would be nice, but I have my doubts. Should we all have equally precise, equally up-to-the-second information on road conditions, the odds are that we'll all respond in similar ways. As we move in unison to avoid one bottleneck, we'll create a new bottleneck. We may come to look back fondly on the days when information was less uniformly distributed.

That's the problem with transparency. When we all know what everyone else knows, it becomes ever harder to escape the pack. Just ask the surfers who dedicate themselves to scouting out the best waves. They used to be able to keep their favorite spots secret, riding their boards in relative solitude. But in recent months people have begun putting up dozens of video cameras, known as "surf cams," along remote shorelines and streaming the pictures over the net. Thanks to the cameras, once-secluded waters are now crowded with hordes of novice surfers. That's led to an outbreak of "surf cam rage." Hard-core surfers are smashing any cameras they find in the hope that they might be able to turn the tide of transparency.

The vandalism may feel good, but it's in vain. For every surf cam broken, a few more go up in its place. Everything that can be streamed will be streamed.

There's a lot to be said for the quick access to information that the internet allows. Some walls that deserve to be torn down are being torn down. At the same time, transparency is erasing the advantages that once went to the intrepid, the dogged, and the resourceful. The ballsy surfer who found the perfect wave off an undiscovered stretch of beach is being elbowed out by the lazy masses that can discover the same wave with just a few mouse clicks. The industrious commuter who studied maps and scouted byways to find shortcuts to work finds herself stuck in a jam with the GPS-enabled horde.

You have to wonder whether, as all that was foggy is rendered clear, the bolder among us will lose the desire to strike out for undiscovered territory. What's the point when every secret becomes, in an internet second, common knowledge?

GILLIGAN'S WEB

May 14, 2008

DESPITE THE BEST EFFORTS of the deletionists, the true glory of Wikipedia continues to lie in the obscure, the arcane, and the ephemeral. Nowhere else will you find such lovingly detailed descriptions of cartoon characters, video games, obsolete software languages, and the workings of machines that exist only in science fiction. Whatever else it may be, Wikipedia is a monument to the obsessive-compulsive fact-mongering of the adolescent male. Never has sexual sublimation been quite so wordy.

My favorite example is the Wik's wonderfully panoramic coverage of the sixties sitcom *Gilligan's Island*. Not only is there an exhaustive entry for the show itself, but there are separate articles for each of the castaways— Gilligan, the Skipper, the Professor, Mary Ann, Ginger, Thurston Howell III, and Eunice "Lovey" Wentworth Howell—as well as the actors who played the roles, the ill-fated SS *Minnow*, and even the subsequent TV movies that were based on the show, including the 1981 classic *The Harlem Globetrotters on Gilligan's Island*. Best of all is the annotated list of all ninety-eight of the episodes in the series, which includes a color-coded guide to "visitors, animals, dreams, and bamboo inventions."

It goes deeper than Wikipedia. *Gilligan's Island* has been a great motivator of user-generated content across the breadth of the web. There's a vast corpus of digital fan fiction that imagines the seven characters in all kinds of dramatic and erotic situations. There are lovingly crafted YouTube videos and Facebook pages dedicated to exploring the eternal question: "Mary Ann or Ginger?" In fact, if I were called in to rename Web 2.0, I think I'd call it Gilligan's Web, if only to underscore the symbiosis between the pop-culture artifacts of the mass media and so much of the user-generated content found online.

So imagine my bewilderment when, a few days ago, I read a transcript of a recent speech that new-media scholar Clay Shirky gave to a Web 2.0 conference in which he argued that *Gilligan's Island* and Web 2.0 are actually opposing forces in the grand sweep of history. Is Professor Shirky surfing a different web than the rest of us?

To Shirky, the TV sitcom, as exemplified by *Gilligan's Island*, was "the critical technology for the 20th century." Why? Because it sucked up all the spare time that people suddenly had on their hands in the decades after the second world war. The sitcom "essentially functioned as a kind of cognitive heat sink, dissipating thinking that might otherwise have built up and caused society to overheat." I'm not exactly sure what Shirky means when he speaks of society overheating, but, anyway, it wasn't until the arrival of the World Wide Web and its "architecture of participation" that we suddenly gained the capacity to do something productive with our "cognitive surplus," like edit Wikipedia articles or play the character of an elf in a World of Warcraft clan. Writes Shirky:

> Did you ever see that episode of *Gilligan's Island* where they almost get off the island and then Gilligan messes up and then they don't? I saw that one. I saw that one a lot when I was growing up. And every half-hour that I watched that was a half an hour I wasn't posting at my blog or editing Wikipedia or contributing to a mailing list. Now I had an ironclad excuse for not doing those things, which is none of those things existed then. I was forced into the channel of media the way it was because it was the only option. Now it's not, and that's the big surprise. However lousy it is to sit in your basement and pretend to be an elf, I can tell you from personal experience it's worse to sit in your basement and try to figure if Ginger or Mary Ann is cuter.

What Shirky is doing here is repackaging the liberation mythology that has long characterized the more utopian writings about the net. That mythology draws a sharp distinction between our lives before the coming of the web (BW) and our lives after the web's blessed advent (AW). In the dark BW years, we were passive couch potatoes who were, in Shirky's words, "forced into the channel of media the way it was because it was the only option." We were driftwood, going with whatever flow "the media" imposed on us. We were all trapped in Shirky's musty cellar. The web, the myth continues, emancipated us. No longer were we forced into the channel of passive consumption. We could "participate." We could "share." We could "produce." When we turned our necks from the TV screen to the computer screen, we slipped the noose.

I think we'd all agree that the web is changing the structure of media, and that's going to have important ramifications. Some will be good, some bad, and the way they'll all shake out remains unknown. But what about Shirky's idea that in the BW years we were unable to do anything "interesting" with our "cognitive surplus"—that the only option was watching TV? That is horseshit. It may well be that Clay Shirky spent all his time pre-1990 watching sitcoms in his cellar (though I very much doubt it), but I was also alive in those benighted years, and I seem to remember a whole lot more going on.

Did my friends and I watch *Gilligan's Island*? You bet we did—and thoroughly enjoyed it (though with a bit more ironic distance than Shirky allows). Watching sitcoms and the other junk served up through the boob tube was part of our lives. But it was not the center of our lives. Most of the people I knew were doing a whole lot of "participating," "producing," and "sharing," and, to boot, they were doing it not only in the symbolic sphere of media but in the actual physical world as well. They were shooting eight-millimeter films, playing drums and guitars and saxophones in bands, composing songs, writing poems and stories, painting pictures, taking and developing photographs, drawing comics, souping up cars, constructing elaborate model railroads in elaborate model landscapes, reading books and watching movies and listening to records and discussing them deep into the night, experimenting with mind-altering substances, volunteering in political campaigns, protesting for various causes, and on and on and on. I'm sorry, but nobody other than the odd loser was stuck, like some pathetic shred of waterborne trash, in a single media-regulated channel.

Tom Slee, in a review of Shirky's new book *Here Comes Everybody*, strips some of the bright varnish from the net's liberation mythology. In the book, Shirky describes, with intelligence and clarity, the social and economic dynamics of virtual communities. But he also, as Slee notes, indulges his enthusiasm for the web in a way that draws, once again, an overly bright line between BW and AW:

> Clay looks at the internet and sees lots of groups forming . . . and he concludes that the world is alight with a new groupiness, the likes of which we have never seen. . . . While Clay is telling us all about the use of digital technology to spark innovative forms of protest in Belarus, which is a fascinating story, we really need . . . to ask why, with all these group-forming tools at our disposal and despite the documented disillusionment with the war in Iraq, there is so little coherent protest happening compared to previous wars? Is it really the case that society now is becoming, thanks to the internet, more democratic, more collaborative, and more coopera- tive than before? I am not convinced.

As Slee suggests, the liberation mythology evaporates when you look at history. It's worth remembering that *Gilligan's Island* origi- nally ran on television from 1964 through 1967, a period noteworthy not for its social passivity but for its social activism. These were years of great cultural and artistic exploration and inventiveness, and they were also years of widespread protest, when people organized into very large—and very real—groups within the civil rights movement, the antiwar movement, the feminist movement, the black power move- ment, the psychedelic movement, and all sorts of other movements. People weren't in their basements; they were in the streets.

If everyone was so enervated by *Gilligan's Island* and other televised piffle, how exactly do you explain 1966 or 1967 or 1968? The answer is: You don't. Indeed, once you begin contrasting 1968 with 2008, you might even find yourself thinking that, on balance, the web is not an engine for social activism but an engine for social passivity. You might even suggest that the web funnels our urges for "participation" and "sharing" into politically and commercially acceptable channels—that it turns us into play-actors, make-believe elves in make-believe clans.

As for the larger question: Mary Ann.

COMPLETE CONTROL

September 3, 2008

THERE'S AN EXHILARATING MOMENT in the middle of the Clash's "Complete Control" when, during Mick Jones's skittering, anarchic solo, Joe Strummer screams, "You're my guitar hero!" He's kidding, but he's not. "Complete Control" is, along with the Sex Pistols' "Anarchy in the UK" and X-Ray Spex's "Oh Bondage Up Yours!," one of the purest distillations of the punk ethos of regeneration through degeneration. The song, ostensibly a broadside against record company power, is a club the Clash wields to smash every constraint in sight, even its own hold over its fans. It accelerates the machinery of pop until the machine disintegrates, and Strummer's scream is the song's climax, maybe the climax of the entire punk antimovement. Strummer, by turning himself into the subject as well as the object of audience worship, frees the fan from his prescribed role as consumer, subverts fandom by turning fandom into an act of subversion. The band is the fan, and the fan is the band, and both stand as one outside and beyond the producer-consumer dynamic that would contain them.

"You're my guitar hero!"

It's a fleeting act, as Strummer well understood. When the song ends so does its spell. We're "in control"—or, more exactly, "out of control"—for three minutes. Then the prescribed identities reassert themselves. When the single came out, late in 1977, my friends and I would play it over and over again, renewing its promise and its illusion. We wanted to be uncontrollable. But we were smart enough to realize that listening to "Complete Control" was also, and already, an act of nostalgia. The song was, as Jon Savage wrote in *England's Dreaming*, "a hymn to Punk autonomy at the moment of its eclipse."

Strummer died in 2002, sparing him the gut-shot of seeing the song included in the new release of the video game Guitar Hero. In the game's color-coded cartoon world, "Complete Control" becomes complete parody. It's the perfect subversion of the Clash's subversion, anarchy turned into routine, with a score-keeping mechanism. Now when Strummer yells "You're my guitar hero!," it's an act of advertising, the cynical come-on of a hawker. Strummer's scream becomes a moment not of mutual liberation but of deep creepiness. The ironies are piled so high that the only way out is to pretend they're not there, to tell yourself that "Complete Control" was never anything but a pop song.

The popularity of Guitar Hero puzzles the social critic Rob Horning: "If you want a more interactive way to enjoy music, why not dance, or play air guitar? Or better yet, if holding a guitar appeals to you, why not try actually learning how to play? For the cost of an Xbox and the Guitar Hero game, you can get yourself a pretty good guitar." Horning, apparently, doesn't quite get the point of "prosumerism," the widely worshipped transformation of creativity into an act of consumerism. Prosumerism's reputed joys are lost on him. He continues: "I can't help but feel that Guitar Hero (much like Twitter) would have been utterly incomprehensible to earlier generations, that it is a symptom of some larger social refusal to embrace difficulty."

In explaining the way that trivial, if diverting, pursuits like Guitar Hero provide an easy alternative to meaningful work, Horning draws on the writing of political theorist Jon Elster. In his 1986 book *An Introduction to Karl Marx*, Elster used a simple example to illustrate the psychic difference between the hard work of developing talent and the easy work of consuming stuff:

Compare playing the piano with eating lamb chops. The first time one practices the piano it is difficult, even painfully so. By contrast, most people enjoy lamb chops the first time they eat them. Over time, however, these patterns are reversed. Playing the piano becomes increasingly more rewarding, whereas the taste for lamb chops becomes satiated and jaded with repeated, frequent consumption.

Elster then made a broader point:

> Activities of self-realization are subject to increasing marginal utility: They become more enjoyable the more one has already engaged in them. Exactly the opposite is true of consumption. To derive sustained pleasure from consumption, diversity is essential. Diversity, on the other hand, is an obstacle to successful self-realization, as it prevents one from getting into the later and more rewarding stages.

"Consumerism," comments Horning, "keeps us well supplied with stuff and seems to enrich our identities by allowing us to become familiar with a wide range of phenomena—a process that the internet has accelerated immeasurably. . . . But this comes at the expense with developing any sense of mastery of anything, eroding over time the sense that mastery is possible, or worth pursuing."

Distraction is the permanent end state of the perfected consumer, not least because distraction is a state that is eminently programmable. To buy a guitar is to open possibilities. To buy Guitar Hero is to close them. A commenter on Horning's article writes, "To me, the radical move that Guitar Hero makes is to turn music into an objectively measurable activity that is more amenable to our Protestant work ethic. It brings the corporation's focus on quantitative performance indicators to the domain of music, displacing the usual mode of subjective enjoyment."

Who's in control?

EVERYTHING THAT DIGITIZES MUST CONVERGE

October 19, 2008

"A CENTRIPETAL FORCE," WROTE Isaac Newton, "is that by which bodies are drawn or impelled, or any way tend, towards a point as to a center." That's a pretty good description of what's been going on with the web.

When I started blogging, back in 2005, I would visit Technorati, the then-popular blog search engine, several times a day, both to monitor mentions of my own blog and to track discussions on subjects I was interested in writing about. But over the last year or so my blog-searching behavior has changed. I started using Google Blog Search to supplement Technorati, and then, without even thinking about it really, I began using Google Blog Search exclusively. At this point, I can't even remember the last time I visited the Technorati site. Honestly, I don't even know if it's still around.

Technorati's technical glitches were part of the reason for the change in my behavior. Even though Technorati offered more precise tools for searching the blogosphere, it was often slow to return results, or it would just fail outright. When it came to handling large amounts of traffic, Technorati couldn't compete with Google's resources. But it wasn't just a matter of responsiveness and reliability. As a web-services conglomerate, Google made it easy to enter one keyword and then do a series of different searches from its site. By clicking on the links to various search engines that Google conveniently arrays across the top of every results page, I could search the web, then search news stories, then search blogs. Google offered the path of least resistance, and I took it.

I thought of this today as I read a report that people seem to be abandoning Bloglines, the online feed reader, and that many of them are coming to use Google Reader instead. The impetus, again, seems to be a mix of frustration with Bloglines's glitches and the availability of a decent and convenient alternative operated by the giant Google.

During the 1990s, when the World Wide Web was new, it exerted a strong centrifugal force. It pulled us out of the orbit of big media outlets and sent us skittering to the outskirts of the info-universe. Early web directories like Yahoo and early search engines like AltaVista, whatever their shortcomings (perhaps *because* of their shortcomings), led us to personal web pages and other small, obscure, and often oddball sources of information. The earliest web loggers, too, took pride in ferreting out and publicizing far-flung sites. And, of course, the big media outlets were slow to move to the web, so their gravitational fields remained weak or nonexistent online. For a time, to bring my metaphor down to earth, the web had no mainstream; there were just brooks and creeks and rills, and the occasional beaver pond.

Those were the days when you could look around and convince yourself that the web would always be resistant to centralization. But that view was an illusion even then. The counterforce to the web's centrifugal force—the centripetal force that would draw us back toward big, central information stores—was building. Hyperlinks were creating feedback loops that served to amplify the popularity of popular sites, feedback loops that would become massively more powerful when modern search engines, like Google, began to rank pages on the basis of links and traffic and other measures of popularity. Navigational tools that used to emphasize ephemera began to filter it out. Roads out began to curve back in.

At the same time, and for related reasons, size began to matter. A lot. Big media outlets moved online, creating vast, enticing pools of branded content. Search engines and content aggregators expanded explosively, providing them with the money and expertise to create technical advantages—in speed and reliability, among other things—that often proved decisive in attracting and holding consumers. The winner-take-all "network effects" of mass communication systems took hold. (Everybody wants to be where everyone else is.) And, of course, people began to demonstrate their innate laziness, retreating from the wilds and following well-trod trails. A Google search may turn up thousands of results, but few of us bother to scroll beyond the top three. When convenience battles curiosity, convenience usually wins.

Wikipedia provides a good example of the self-reinforcing power of the web's centripetal force. The popular online encyclopedia is less the sum of human knowledge than the black hole of human knowledge. A vast exercise in cut-and-paste paraphrasing (it explicitly bans original thinking), Wikipedia first sucks in content from other sites, then it sucks in links, then it sucks in search results, then it sucks in readers. And because it prevents search engines from taking account of its outbound links to the sources of its articles, through the use of "no follow" tags, it reinforces its hegemony over search results. Light comes in but doesn't go out. One of the untold stories of Wikipedia is the way it has siphoned traffic from smaller specialized sites, like the excellent Stanford Encyclopedia of Philosophy, even though those sites often have better information about the topics they cover. Wikipedia articles have become the default external link for many creators of web content, not because Wikipedia is the best source but because it's the best-known source and, generally, it's "good enough." Wikipedia is the lazy man's link, and we're all lazy men, except for those of us who are lazy women.

What Chris Anderson famously described as the long tail of content still exists online, but far from wagging the web-dog, the tail has taken on the look of a vestigial organ. Chop it off, and most people would hardly notice. On the net as off, things gravitate toward large objects. The center holds.

RESURRECTION

February 16, 2009

THE SINGULARITY—THAT much-anticipated moment when artificial intelligence leaps ahead of human intelligence, rendering man immortal at the instant of his obsolescence—has been called "the rapture of the geeks." But to Ray Kurzweil, the most famous of the Singularitarians, it's no joke. In an interview in *Rolling Stone*, Kurzweil describes how, in the wake of the Singularity, it will be possible not only to preserve the living for eternity (by uploading their minds into computers) but to resurrect the dead (by reassembling the information that formed their vital essence). Life is data, and data never die.

Kurzweil seems pretty certain about what the future holds. He predicts that advances in artificial intelligence and nanotechnology will soon allow the construction of a revivification machine, a kind of Polaroid camera that produces not an image of a person but the living being. He looks forward in particular to reuniting with his beloved father, Fredric, who died in 1970, when Ray was just twenty-two.

In a soft voice, he explains how the resurrection would work. "We can find some of his DNA around his grave site—that's a lot of information right there," he says. "The AI will send down some nanobots and get some bone or teeth and extract some DNA and put it all together. Then they'll get some information from my brain and anyone else who still remembers him." . . . To provide the nanobots with even more information, Kurzweil is safeguarding the boxes of his dad's mementos, so the artificial intelligence has as much data as possible from which to reconstruct him.

There's a real poignancy to Kurzweil's dream of bringing his dad back to life by weaving strands of DNA with strands of memory. I can imagine a novel—by Edgar Allan Poe, maybe, or Joyce Carol Oates—built around this otherworldly yet altogether American yearning. Death makes strange even the most logical of minds.

ROCK-BY-NUMBER

August 18, 2009

THE RELEASE NEXT MONTH of The Beatles: Rock Band, the latest edition of the popular video game series, is shaping up to be the cultural event of the year, if not the millennium. The game was the subject of an epic article by Daniel Radosh in Sunday's *New York Times*, which featured comments from Paul and Ringo as well as John's widow, George's widow, and George Martin's son. Apple Corps, reports Radosh, hopes the game "will be the most deeply immersive way ever of experiencing the music and the mythology of the Beatles." The CEO of Harmonix Music Systems, the company making the game, says, "We're on the precipice of a culture shift around how the mass market experiences music." Adds Radosh: "Playing music games requires an intense focus on the separate elements of a song, which leads to a greater intuitive knowledge of musical composition."

In the 1950s, paint-by-number kits were all the rage. Everyone became an artist, diligently filling in the numbered spaces on prefabricated canvases with specified colors. Today, even as we celebrate the contrivances of Rock Band, we look down our noses at those lowbrow paint-by-number kits. Yet I'm sure that somebody back in the fifties wrote about how paint-by-number "requires an intense focus on the separate elements of a painting, which leads to a greater intuitive knowledge of artistic composition." I'm sure it was thought that paint-by-number liberated people from being passive observers of art, that it allowed them to participate, in a deeply immersive way, in the act of painting.

We shouldn't be too hard on our fads. They become fads because they're fun. Still, a fad will always tell us something important about the times in which it takes hold. "Paint-by-number," wrote Brennen

Jensen in a 2001 *City Paper* article, "is all about conformity. Indeed, there is perhaps no greater metaphor for America's *Leave It to Beaver*, I Like Ike, *Man in the Gray Flannel Suit* 1950s than this 'digital art' craze that roared through the decade, promising neophyte brush-wielders 'a beautiful oil painting the first time you try!' " Painting, when done by number, "is rendered rote—a matter of manual dexterity, not inspiration."

Rock Band is the aural equivalent of paint-by-number. It's musicianship-by-number, once again substituting dexterity for inspiration. It's also a fad. Ten or so years from now, we'll look back on the game with a mixture of nostalgia and embarrassment.

But, like paint-by-number, Rock Band is also a metaphor. As you, the targeted mass-market player, proceed through The Beatles game, you time travel through a waxworks rendering of the sixties. You go from the go-go soundstage of the *Ed Sullivan Show* to the trippy mindscapes of psychedelia to the flowerchild fields of the hippies. It's a paint-by-number version of history, one that co-opts the anticonsumerist spirit of the counterculture in order to betray it. The Beatles: Rock Band is an exercise in simulation, or, as we say today, virtualization. Unlike the seditious simulation of, say, Warhol's pop art, which by draining the Campbell's soup can of its comforting contents set the consumer at a discomfiting distance from the act of consumption, the simulation of Rock Band is cynical. It requires of us only that we follow instructions, play by the rules. Where paint-by-number was conformist, Rock Band is calculating.

RAISING THE
VIRTUAL CHILD

February 21, 2010

I REGULARLY RECEIVE EMAILS and texts from panicked parents worried that they may be failing in what has become the central challenge of modern child-rearing: ensuring that kids grow up well adapted to the online environment. These parents are concerned—and rightly so—that their offspring will be at a disadvantage in the digital milieu in which we all increasingly live, work, love, and compete for the small bits of attention that, in the aggregate, define the success, or failure, of our days. If maladapted to virtuality, these moms and dads understand, their progeny will end up socially ostracized, with few friends and even fewer followers. "Can we even be said to be alive," one agitated young mother wrote me, "if our status updates go unread?" The answer, of course, is no. In the data-stuffed world of real-time messaging, the absence of interactive stimuli, even for brief periods, may result in a state of reflective passivity indistinguishable from nonexistence. On a more practical level, a lack of online skills is sure to constrain a young person's long-term job prospects. At best, she will be fated to spend her days involved in some form of manual labor, possibly even working out of doors with severely limited access to screens. At worst, she will have to find a non-tenure-track position in academia.

Fortunately, raising what I call the virtual child is not difficult. The newborn human infant, after all, leads a purely real-time existence, immersed in a "stream" of alerts and stimuli. As long as the child is kept in the crosscurrents of the messaging stream *from the moment of parturition*—the biological womb replaced immediately with the wi-fi womb—adaptation to virtuality will likely be seamless and complete. It is only when a sense that time may consist of something other than the immediate moment is allowed to impinge on the child's conscious-ness that maladaptation becomes a real possibility. Hence, the most pressing job for the parent is to ensure that the virtual child is kept in a device-rich networked environment at all times.

It is also essential that the virtual child never be allowed to run a cognitive surplus. His or her mental accounts must always be kept in perfect balance, with each synaptic firing being immediately deployed for a well-defined chore, preferably involving the manipulation of sym-bols on a computer screen in a collaborative social-production exercise. If cognitive cycles are allowed to go to waste, the child may drift into an introspective "dream state" outside the flow of the digital stream. It is wise to ensure that your iPhone is well populated with apps suitable for children, as this will provide a useful backup should your child break, lose, or otherwise be separated from her own network-enabled devices. Printed books should in general be avoided, as they also tend to promote an introspective dream state, though multifunctional devices that include e-reading apps, such as Apple's iPad, are permissible.

The out-of-doors poses particular problems for the virtual child, as nature has in the past earned a reputation for inspiring states of intro-spectiveness and even contemplativeness in impressionable young peo-ple. (Some psychologists suggest that just looking out a window may be dangerous to the mental health of the virtual child.) Sometimes it is simply impractical to keep a tyke from interacting with the natural world. At such moments, it is all the more important that the child be outfitted with portable electronic devices, including music players, smartphones, and gaming instruments, in order to ensure no break in the digital stream. If you are not able to physically accompany your child on expeditions into the natural world, it is a good idea to send text messages to your child every few minutes just to be on the safe side.

The challenges of keeping your child immersed in an online envi-ronment can be trying, but remember: History is on your side. Virtu-ality becomes increasingly ubiquitous with each passing day. It is also important to remember that one of the great joys of modern parent-hood is documenting your virtual infant's or toddler's special moments through texts, tweets, posts, uploaded photos, and YouTube clips. The virtual child presents ideal messaging-fodder for the virtual parent.

Navigating the real-time messaging stream is a journey that you and your child take together. Every moment is unique because every moment is disconnected from both the one that precedes it and the one that follows it. Virtuality is a state of perpetual renewal. The joy of infancy continues forever.

THE IPAD LUDDITES

April 7, 2010

IS IT POSSIBLE FOR a Geek God to also be a Luddite? That was the question that popped into my head as I read Cory Doctorow's impassioned anti-iPad diatribe at Boing Boing. The device that Apple calls "magical" and "revolutionary" is, to Doctorow, a counterrevolutionary contraption conjured up through the black magic of the evil wizards at One Infinite Loop. The locked-down, self-contained design of the iPad—nary a USB port in sight, and don't even think about loading an app that hasn't been blessed by Steve Jobs—manifests "a palpable contempt for the owner," writes Doctorow. You can't fiddle with the damn thing.

> The original Apple II came with schematics for the circuit boards, and birthed a generation of hardware and software hackers who upended the world for the better. . . . Buying an iPad for your kids isn't a means of jump-starting the realization that the world is yours to take apart and reassemble; it's a way of telling your offspring that even changing the batteries is something you have to leave to the professionals.

Doctorow is not the only nerd who's uncomfortable with Apple's transformation of the hacktastic PC into a sleek, slick, sterile appliance. Many have accused Apple of removing from the personal computer not only its openness and open-endedness but also what Jonathan Zittrain, founder of Harvard's Berkman Center for Internet & Society, calls its "generativity"—its capacity for encouraging and abetting creative work. In criticizing the closed nature of the iPhone, from which the iPad borrows its operating system, Zittrain, like Doctorow, invoked the ancient, beloved Apple II: "a clean slate, a device built—boldly—with no specific tasks in mind."

What these folks are ranting against is progress—more precisely, progress that goes down a path they don't approve of. They want progress to follow their own ideological bent, and when it takes a turn they don't like they start grumbling like fogies, yearning for the days of their idealized Apple IIs, when men were men and computers were computers.

If Ned Ludd had been a blogger, he would have written a post similar to Doctorow's about those newfangled, locked-down mechanical looms that distance the weaver from the machine's workings, requiring the weaver to follow the programs devised by the loom manufacturer. The design of the mechanical loom, Ned would have told us, exhibits a palpable contempt for the user. It takes the generativity out of weaving.

And Ned would have been right.

I have a lot of sympathy for the point of view expressed by Doctorow, Zittrain, and other defenders of the PC faith. The iPad, for all its glitzy technical virtuosity, does feel like a step backward from the Apple II and its progeny. Hell, I still haven't gotten over Apple's removal of analog RCA plugs for audio and video input and output from the back of its Macs. Give me a beige box with easily accessible innards, a big rack of RAM, and a dozen or so ports, and I'm good.

But progress doesn't give a crap about what I want. While progress may at times be spurred by the hobbyist, it doesn't share the hobbyist's ethos. One of the keynotes of technological advance is its tendency, as it refines a tool, to remove real human agency from the tool's workings. In its place, we get an abstraction of human agency that represents the general desires of the masses as deciphered, or imposed, by the manufacturer and the marketer. Indeed, what tends to distinguish the advanced device from the primitive device is the absence of generativity. It's worth remembering that the earliest radios were broadcasting devices as well as listening devices and that the earliest phonographs could be used for recording as well as playback. But as these machines evolved, along with the media systems in which they became embedded, they turned into streamlined, single-purpose entertainment boxes, suitable for living rooms. What the iPad Luddites fear—the divergence of the creative tool from the mass-market device—has happened over and over again, usually without much, if any, resistance on the part of buyers.

Progress may, for a time, intersect with one's own personal beliefs, and during that period one will become a gung-ho technological progressive. But that's just coincidence. In the end, progress is oblivious to anyone's beliefs or yearnings. Those who think of themselves as great fans of progress, of technology's inexorable march forward, will change their tune as soon as progress destroys something they care deeply about. Passion turns us all into primitivists.

NOWNESS

June 8, 2010

"THOUGHT WILL SPREAD ACROSS the world with the rapidity of light, instantly conceived, instantly written, instantly understood. It will blanket the earth from one pole to the other—sudden, instantaneous, burning with the fervor of the soul from which it burst forth." That's from a Google press release, announcing a big new investment in fiber optics.

I'm lying. The words were written in 1831 by the French poet and politician Alphonse de Lamartine, and what Lamartine heralded was the arrival of the daily newspaper. Journalism would soon become "the whole of human thought," he predicted. Books, unable to compete with the immediacy of morning and evening papers, were doomed: "Thought will not have time to ripen, to accumulate into the form of a book—the book will arrive too late. The only book possible from today is a newspaper."

Lamartine's prediction of the book's demise didn't pan out. But he was a prophet all the same. The story of media has for the last two centuries been a story of the pursuit of ever greater immediacy. From broadsheet to telegram to radio broadcast to TV show to blog to microblog, we have been hell-bent on ratcheting up the velocity of information. It's funny, kind of, that Lamartine's agent of immediacy is immediacy's latest victim. The newspaper arrives too late.

"Ripeness is all," said Gloucester's boy in *King Lear*, and we were inclined to believe him. No more. Ripeness is nothing. Ripeness is for the landfill. Nowness is all.

CHARLIE BIT MY
COGNITIVE SURPLUS

August 3, 2010

"You can say this for the technological revolution; it's cut way down on television." So writes Rebecca Christian in a column for the *Telegraph Herald* in Dubuque. She's not alone in thinking that the time we devote to the web is reducing the time we spend watching the tube. It's a common assumption. And it's wrong. Despite the rise of digital media, Americans are watching more TV than ever.

The Nielsen Company has been tracking media use for decades, and it reported last year that in the first quarter of 2009, the time Americans spent watching TV hit its highest level ever—the average person was watching 156 hours and 24 minutes of TV a month. Now, Nielsen has come out with an update for the first quarter of 2010. Once again, TV viewing has hit a record, with the average American spending 158 hours and 25 minutes a month in front of a television set. Although two-thirds of Americans now have broadband internet access at home, TV viewing continues its seemingly inexorable rise.

And the Nielsen TV numbers actually understate our consumption of video programming, because the time we spend viewing video on our computers and cell phones is also going up. The average American with internet access is now watching 3 hours and 10 minutes of video on net-connected computers every month, Nielsen reports, and the average American with a video-capable cell phone is watching an additional 3 hours and 37 minutes of video on the phone every month. Not surprisingly, expanding people's access to video programming increases their consumption of that programming.

What about the young? Surely, so-called digital natives are watching less TV, right? Nope. The young, too, continue to ratchet up their TV viewing. A recent study of media habits by Deloitte showed, in fact, that over the past year people in the fourteen to twenty-six age bracket increased their TV viewing by a greater percentage than any other age group. An extensive Kaiser Family Foundation study released earlier this year found that while young people appear to be spending a little less time in front of TV sets today than they did five years ago, that decline is offset by increased viewing of television programming on computers, phones, and tablets. Overall, "the proliferation of new ways to consume TV content has actually led to an *increase* of 38 minutes of daily TV consumption" by the young, reports Kaiser. Nielsen, too, finds that TV viewing continues to rise among children, teens, and young adults.

What about the rise of amateur media production, abetted by sites like YouTube? That trend, at least, must be shifting us away from media consumption. Wrong again. As Bradley Bloch explained in a recent Huffington Post article, the ease with which amateur media productions can be distributed online actually has the paradoxical effect of increasing people's media consumption far more than it increases their media production. "Even if we count posting a LOLcat as a creative act," observes Bloch, "there are many more people looking at LOL-cats than there are creating them." Bloch runs the numbers on one oft-viewed YouTube entertainment: "One of the most popular videos on YouTube, 'Charlie bit my finger—again!,' depicting a boy sticking his fingers in his little brother's mouth, has been viewed 211 million times. Something that took 56 seconds to create—and which was only intended to be seen by the boys' godfather—has sucked up the equivalent of 1600 people working 40 hours a week for a year." By giving us easy and free access to millions of short-form video programs, the web allows us to cram ever more video viewing into the nooks and crannies of our daily lives.

To give an honest accounting of the effects of the net on media consumption, you need to add the amount of time that people spend consuming web media to the amount of time they already spend consuming TV and other traditional media. Once you do that, it becomes clear that the net has not reduced the time devoted to imbibing media but increased it, a lot. The web, in other words, marks a continuation of a long-term cultural trend, not a reversal of it. The difference is, you no longer need a couch to be a couch potato. With smartphone in hand, you can be a spud wherever you go.

MAKING SHARING SAFE
FOR CAPITALISTS

November 8, 2010

"I AM NOT A COMMUNIST," declared author-entrepreneur Steven Johnson in a recent column in the business pages of the *New York Times*. Johnson made his disclaimer in the course of celebrating the creativity of "open networks," the groups of volunteers who gather on the net to share ideas and produce digital goods of one stripe or another. Because they exist outside the marketplace and don't operate in response to the profit motive, one might think that such social-production collectives would represent a threat to traditional markets. What could be more subversive to consumer capitalism than a mass movement of people working without pay to create free stuff for other people? But capitalists needn't worry. The innovations of the unpaid, web-enabled masses may be "conceived in nonmarket environments," writes Johnson, but they create "new platforms" that "support commercial ventures." The net allows the efforts of volunteers to be turned into the raw material for profit-making companies.

Johnson's view is typical of many of the web's most enthusiastic promoters, the Corporate Communalists who feel compelled to distance themselves from, if not ignore entirely, the more radical implications of the trends they celebrate. In a new book with a vaguely Marxist title, *What's Mine Is Yours*, business consultants Rachel Botsman and Roo Rogers begin by describing the onset of what sounds like an anti-market revolution. "The convergence of social networks, a renewed belief in the importance of community, pressing environmental concerns, and cost consciousness," they write, "are moving us away from the old, top-heavy, centralized, and controlled forms of consumerism toward one of sharing, aggregation, openness, and cooperation." We are at a moment of transition from "the twentieth century of hyper-consumption," when "we were defined by credit, advertising, and what we owned," to "the twenty-first century of Collaborative Consumption," when "we will be defined by reputation, by community, and by what we can access and how we share and what we give away."

But, having raised the specter of an anticonsumerist insurrection, Botsman and Rogers immediately defuse it. Like Johnson, they turn out to be more interested in the way online sharing feeds into profit-making ventures. "Perhaps what is most exciting about Collaborative Consumption," they write, with charming naiveté, "is that it fulfills the hardened expectations on both sides of the socialist and capitalist ideological spectrum without being an ideology in itself." In fact, "for the most part, the people participating in Collaborative Consumption are not Pollyannaish do-gooders and still very much believe in the principles of capitalist markets and self-interest. . . . Collaborative Consumption is by no means antibusiness, antiproduct or anticonsumer." Whew!

Botsman and Rogers are more interested in co-opting anticonsumerist energies than unleashing them. A similar tension, between revolutionary rhetoric and counterrevolutionary message, runs through the popular "wikinomics" writings of consultants Don Tapscott and Anthony Williams. In their new book, *Macrowikinomics*, they again promote the net as, to quote from Tom Slee's trenchant review, "a revolutionary force for change, carrying us to a radically different future." And yet, as Slee goes on to point out, the blurbs on the back of the book come from a who's who of big company CEOs. The populist revolution that Tapscott and Williams describe is one that bears, explicitly, the imprimatur of Davos billionaires. For these authors, too, the ultimate promise of open networks lies in providing new opportunities, or "platforms," for profiteers.

What most characterizes today's web revolutionaries is their rigorously apolitical and ahistorical perspectives—their fear of actually being revolutionary. To them, the disruptions of the web end in a reinforcement of the status quo. There's nothing wrong with that view,

I suppose—these are all writers who court business audiences—but their writings do testify to just how far we've come from the idealism of the early days of cyberspace. Back then, online communities were proudly anticommercial, and the net's free exchanges stood in opposition to what John Perry Barlow, in his "Declaration of the Independence of Cyberspace," dismissively termed "the Industrial World." By encouraging us to think of sharing as Collaborative Consumption, the technologies of the web now look like they will have, as their true legacy, the spread of market forces into the most intimate spheres of life.

THE QUALITY OF ALLUSION IS NOT GOOGLE

January 15, 2011

ADAM KIRSCH, THE USUALLY deft critic, goes agley in a new *Wall Street Journal* column, "Literary Allusion in the Age of Google." The piece begins well, as things that go agley so often do. Kirsch describes how the art of allusion has waned along with the reading of the classics. As one's personal store of literary knowledge shrinks, so too does one's capacity for allusiveness. But Kirsch also believes that, as our cultural kitty has come to resemble Mother Hubbard's cupboard, the making of a literary allusion has turned into an exercise in elitism. Rather than connecting writer and reader, it separates them. "It doesn't matter whether you slip in 'April is the cruelest month,' or 'To be or not to be,' or even 'The Lord is my shepherd,'" Kirsch argues; "there's a good chance that at least some readers won't know what you're quoting, or that you're quoting at all."

No need to fret, though. Google to the rescue. By allowing even the most literarily challenged reader to find the source of "any quotation in any language" with just a keystroke or two, the search engine is making the world safe again for allusion. "When T. S. Eliot dropped outlandish Sanskrit and French and Latin allusions into 'The Waste Land,' he had to include notes to the poem," writes Kirsch. "Today, no poet could outwit any reader who has an Internet connection." The upshot: Allusions have become "more democratic" and "more generous." Literature, like the world, has been flattened.

It's a dicey proposition, these days, to take issue with a cultural democratizer, a leveler of playing fields, but there are big problems with Kirsch's assessment, and they stem from his desire to see "allusion" as being synonymous with "citation" or "quotation." An allusion is not a citation. It's not a quotation. It's not a pointer. It's not a shout-out. And it definitely is not a hyperlink. An allusion is a hint, a suggestion, a tease, a wink, sometimes a quiet homage. The reference it contains is implicit rather than explicit. Its essential quality is playfulness; the word "allusion" derives from the Latin verb *alludere*, meaning "to play with" or "to joke around with."

The lovely fuzziness of a literary allusion—the way it blurs the line between speaker and source—is the essence of its art. It's also what makes the allusion an endangered species in the Age of Google. A computerized search engine can swiftly parse explicit connections like citations, quotations, and hyperlinks—it feeds on them as a whale feeds on plankton—but it has little sensitivity to more delicate connections, to the implicit, the playful, the covert, the slant. Search engines are literal-minded, not literary-minded. Google's overarching goal is to make culture machine-readable. We've all benefited from its pursuit of that goal, but Google's vast field of vision has a very large blind spot. Much of what's most subtle and valuable in culture—and the allusions of artists fall into this category—is too blurry to be read by machines.

Kirsch says that T. S. Eliot had to append notes to "The Waste Land" in order to enable readers to track down its many allusions. The truth is less clear-cut. The first publications of the poem, in the magazines *The Criterion* and *The Dial*, lacked the notes. The notes only appeared when the poem was published as a book, and Eliot came to rue their addition. The glosses "stimulated the wrong kind of interest among the seekers of sources," he wrote. "I regret having sent so many enquirers off on a wild goose chase after Tarot cards and the Holy Grail." By turning his allusions into mere citations, the notes led readers to see his poem as an intricate intellectual puzzle rather than a profound expression of personal emotion—a confusion that continues to haunt, and hamper, readings of the poem to this day. The beauty of "The Waste Land" lies not in its sources but its music, which is in large measure the music of allusion, of fragments of distant melodies woven into something new. The more you Google "The Waste Land," Eliot would have warned, the less of it you'll hear.

Let's say, to bring in another poet, you're reading Yeats's "Easter 1916," and you reach these lines:

And what if excess of love
Bewildered them till they died?

You would find the poem all the more meaningful, all the more moving, if you caught the allusion to Shelley's "Alastor" ("His strong heart sunk and sickened with excess / Of love . . ."), but the allusion deepens and enriches Yeats's poem whether or not you pick up on it. What matters most is not that you know "Alastor" but that Yeats knows it, and that his reading of the earlier work, and his emotional connection with it, resonates through his own lyric. And since Yeats provides no clue that he's alluding to another work, Google would be no help in tracking down the source of the allusion. A reader who doesn't already have an intimate knowledge of "Alastor" would have no reason to Google the lines.

For the lines to be Google-friendly, the allusion would have to be transformed into a quotation:

And what if "excess of love"
Bewildered them till they died?

or, worse yet, a hyperlink:

And what if <u>excess of love</u>
Bewildered them till they died?

As soon as an allusion is turned into an explicit citation in this way—as soon as it's made fit for the Age of Google—it ceases to be an allusion, and it loses much of its emotional timbre. Distance is inserted between speaker and source. The lovely fuzziness is scrubbed away, the music lost.

In making an allusion, a writer (or a filmmaker, or a painter, or a composer) is not trying to "outwit" the reader (or viewer, or listener), as Kirsch suggests. Art is not a parlor game. Nor is the artist trying to create a secret elitist code that will alienate readers or viewers. An allusion, when well made, is an act of generosity through which an artist shares with the audience a deep attachment with an earlier work or influence. If you see an allusion merely as something to be tracked down, to be Googled, you miss its point and its power. An allusion doesn't become more generous when it's "democratized"; it simply becomes less of an allusion.

My intent here is not to knock Google. It's to point out that there are many ways to view the world and that Google offers only one view, and a limited one at that. One of the dangers we face as we adapt to the Age of Google is that we will all come to see the world through Google goggles, and when I read an article like Kirsch's, with its redefinition of "allusion" into Google-friendly terms, I sense the increasing hegemony of the Google view. It's already becoming common for journalists to tailor headlines and stories to fit the protocols of search engines. Should writers and other artists be tempted to make their allusions a little more explicit, a little more understandable to literal-minded machines, before we know it allusiveness will have been redefined out of existence.

SITUATIONAL OVERLOAD AND AMBIENT OVERLOAD

March 7, 2011

"IT'S NOT INFORMATION OVERLOAD. It's filter failure." That was the theme of an influential talk that Clay Shirky gave at a technology conference in 2008. It's an idea that's easy to like both because it feels intuitively correct and because it's reassuring: Better filters will help reduce information overload, and better filters are things we can actually build. Information overload isn't an inevitable side effect of information abundance. It's a problem that has a solution. So let's roll up our sleeves and start coding.

There was one thing that bugged me, though, about Shirky's idea, and it was this paradox: The quality and speed of our information filters have been improving steadily for a few centuries, and have been improving extraordinarily quickly for the last two decades, and yet our sense of being overloaded with information is stronger than ever. If, as Shirky argues, improved filters will reduce overload, then why haven't they done so up to now? Why don't we feel that information overload is subsiding as a problem rather than getting worse? The reason, I've come to believe, is that Shirky's formulation gets it backward. Better filters don't mitigate information overload; they intensify it. It would be more accurate to say: "It's not information overload. It's filter success."

But let me back up a little, because it's actually more complicated than that. One of the traps we fall into when we talk about information overload is that we're usually talking about two very different things as if they were one thing. Information overload actually takes two forms, which I'll call *situational overload* and *ambient overload*, and they need to be treated separately.

Situational overload is the needle-in-the-haystack problem: You need a particular piece of information—in order to answer a question of one sort or another—and that piece of information is buried in a bunch of other pieces of information. The challenge is to pinpoint the required information, to extract the needle from the haystack, and to do it as quickly as possible. Filters have always been pretty effective at solving the problem of situational overload. The introduction of indexes and concordances—made possible by the earlier invention of alphabetization—helped solve the problem with books. Card catalogs and the Dewey decimal system helped solve the problem with libraries. Train and boat schedules helped solve the problem with transport. The *Readers' Guide to Periodical Literature* helped solve the problem with magazines. And search engines and other computerized navigational and organizational tools have helped solve the problem with online databases.

Whenever a new information medium comes along, we quickly develop good filtering tools that enable us to sort and search the contents of the medium. That's as true today as it's ever been. In general, I think you could make a strong case that, even though the amount of information available to us has exploded in recent years, the problem of situational overload has continued to abate. Yes, there are still frustrating moments when our filters give us the hay instead of the needle, but for most questions most of the time, search engines and other digital filters, or software-based, human-powered filters like email or Twitter, are able to serve up good answers in an eyeblink or two.

Situational overload is not the problem. When we complain about information overload, what we're usually complaining about is ambient overload. This is an altogether different beast. Ambient overload doesn't involve needles in haystacks. It involves haystack-sized piles of needles. We experience ambient overload when we're surrounded by so much information *that is of immediate interest to us* that we feel overwhelmed by the never-ending pressure of trying to keep up with it all. We keep clicking links, keep hitting the refresh key, keep opening new tabs, keep checking email in-boxes and social media feeds, keep scanning Amazon and Netflix recommendations—and yet the pile of interesting information never shrinks.

The cause of situational overload is too much noise. The cause of ambient overload is too much signal.

The great power of modern digital filters lies in their ability to make information that is of inherent interest to us immediately visible to us. The information may take the form of personal messages or updates from friends or colleagues, broadcast messages from experts or celebrities whose opinions or observations we value, headlines and stories from writers or publications we like, alerts about the availability of various other sorts of content on favorite subjects, or suggestions from recommendation engines—but it all shares the quality of being tailored to our particular interests. It's all needles. And modern filters don't just organize that information for us; they push the information at us as alerts, updates, streams. We tend to point to spam as an example of information overload. But spam is just an annoyance. The real source of information overload, at least of the ambient sort, is the stuff we like, the stuff we want. And as filters get better, that's exactly the stuff we get more of.

It's a mistake, in short, to assume that as filters improve they have the effect of reducing the information we have to look at. As today's filters improve, they expand the information we feel compelled to take notice of. Yes, they winnow out the uninteresting stuff (imperfectly), but they deliver a vastly greater supply of interesting stuff. And precisely because the information is of interest to us, we feel pressure to attend to it. As a result, our sense of overload increases. This is not an indictment of modern filters. They're doing precisely what we want them to do: find interesting information and make it visible to us. But it does mean that if we believe that improving the workings of filters will save us from information overload, we're going to be very disappointed. The technology that creates the problem is not going to make the problem go away. If you really want a respite from information overload, pray for filter failure.

GRAND THEFT ATTENTION

April 1, 2011

HAVING RECENTLY COME OFF an Xbox jag, I decided, as an act of penance, to review the latest studies on the cognitive effects of video games. Because gaming has become such a popular pastime so quickly, it has, like television before it, become a focus of psychological and neurological experiments. The research has, on balance, tempered fears that video games would turn players into boggle-eyed, bloody-minded droogs intent on ultraviolence. The evidence suggests that spending a lot of time playing action games—the ones in which you run around killing things before they kill you (there are many variations on that theme)—actually improves certain cognitive functions, such as hand-eye coordination and visual acuity, and can speed up reaction times. In retrospect, these findings shouldn't have come as a surprise. As anyone who has ever played an action game knows, the more you play it, the better you get at it, and getting better at it requires improvements in hand-eye coordination and visual acuity. If scientists had done the same sort of studies on pinball players fifty years ago, they would have probably seen similar results.

But these studies have also come to be interpreted in broader terms. Some popular-science writers draw on them as evidence that the heavy use of digital media—not just video games, but browsing, texting, online multitasking, and so forth—actually makes us "smarter." The foundational text here is Steven Johnson's *Everything Bad Is Good for You*. Johnson draws on an important 2003 study, published in *Nature*, by University of Rochester researchers Shawn Green and Daphne Bavelier, which demonstrated that "10 days of training on an action game is sufficient to increase the capacity of visual attention, its spatial distribution and its temporal resolution." In other words, playing an action game can help you keep track of more visual stimuli more quickly and across a broader field, and these gains may persist even after you walk away from the gaming console. Other studies, carried out both before and after the Green and Bavelier research, generally back up these findings. In his book, Johnson concluded, sweepingly, that video games "were literally making [players] perceive the world more clearly," and he suggested that research on gamers "showed no evidence of reduced attention spans compared to non-gamers."

More recently, the technology reporter Nick Bilton, in his 2010 book *I Live in the Future & Here's How It Works*, also suggested that video-gaming improves attentiveness as well as visual acuity and concluded that "the findings argue for more game playing." The science writer Jonah Lehrer last year argued that video-gaming leads to "significant improvements in performance on various cognitive tasks," including not only "visual perception" but also "sustained attention" and even memory. In her new book *Now You See It*, Cathy Davidson, an English professor at Duke, devotes a chapter to video game research, celebrating a wide array of apparent cognitive benefits, particularly in the area of attentiveness. Quoting Green and Bavelier, Davidson notes, for example, that "game playing greatly increases 'the efficiency with which attention is divided.'"

The message is clear and, for those of us with a fondness for games, reassuring: Fire up the console, grab the controller, and give the old gray matter a workout. The more you play, the smarter you'll get.

If only it were so. The fact is, such broad claims about the cognitive benefits of video games, and by extension other digital media, have always been dubious. They stretch the truth. The mental faculties of attention and memory have many different facets—psychologists and neuroscientists are still a long way from hashing them out—and to the extent that past gaming studies demonstrate improvements in these areas, they relate to gains in the kinds of attention and memory used in the fast-paced processing of a welter of visual stimuli. If you improve your ability to keep track of lots of images flying across a screen, that improvement can be described as an improvement in a type of attentiveness. And if you get better at remembering where you are in a complex fantasy world, that improvement can be described as an improvement in a sort of memory. The improvements may well be real—and that's good news—but they're narrow, and they come with costs. The fact that video games seem to make us more efficient at dividing our attention is great, as long as you're doing a task that requires divided attention (like playing a video game). But if you're actually trying to do something that demands undivided attention, you may find yourself impaired. As UCLA developmental psychologist Patricia Greenfield, one of the earliest researchers on video games, has pointed out, using media that train your brain to scatter attention appears to make you less able to carry out the kinds of deep thinking that require a calm, focused mind. Optimizing for divided attention means suboptimizing for concentrated attention.

Recent studies back up this point. They paint a darker picture of the consequences of heavy video-gaming, particularly when it comes to attentiveness. Far from making us smarter, heavy gaming seems to be associated with attention disorders in the young and, more generally, with a greater tendency toward distractedness and a reduced aptitude for maintaining one's focus and concentration. Playing lots of video games, these studies suggest, does not improve a player's capacity for sustained attention, as Lehrer and others argue. It weakens it.

In a 2010 paper published in the journal *Pediatrics*, Edward Swing and a team of Iowa State psychologists reported on a thirteen-month study of the media habits of some fifteen hundred kids and young adults. It found an association between video games and attention problems, during childhood and on into adulthood. The findings indicate that the correlation between video-gaming and attention disorders is at least equal to and probably greater than the correlation between TV viewing and such disorders. Importantly, the design of the study "rules out the possibility that the association between screen media use and attention problems is merely the result of children with attention problems being especially attracted to screen media."

A 2009 study by a different group of Iowa State researchers, published in *Psychophysiology*, investigated the effects of video-gaming on cognitive control, through experiments with fifty-one young men, both heavy gamers and light gamers. The study indicated that video-gaming has little effect on "reactive" cognitive control—the ability to respond to some event after it happens. But when it comes to "proactive" cognitive control—the ability to plan and adjust one's behavior in advance of an event or stimulus—video-gaming has a significant negative effect. "The negative association between video game experience and proactive cognitive control," the researchers write, "is interesting in the context of recent evidence demonstrating a similar correlation between video game experience and self-reported measures of attention deficits and hyperactivity. Together, these data may indicate that the video game experience is associated with a decrease in the efficiency of proactive cognitive control that supports one's ability to maintain goal-directed action when the environment is not intrinsically engaging." Video gamers, in other words, seem to have a difficult time staying focused on a task that doesn't involve constant incoming stimuli. Their attention wavers.

These findings are consistent with more general studies of media multitasking. In a much-cited 2009 paper in *Proceedings of the National Academy of Sciences*, for example, Stanford researchers showed that heavy media multitaskers have less control over their thoughts than do light multitaskers. The heavy multitaskers "have greater difficulty filtering out irrelevant stimuli from their environment" and are also less able to suppress irrelevant memories from intruding on their work. The heavy multitaskers were actually less efficient at switching between tasks—in other words, they were worse at multitasking itself.

So should people be prevented from playing video games? No. Moderate game-playing probably isn't going to have any significant long-term cognitive consequences, either good or bad. Video-gaming is fun and relaxing, and those are good things. Besides, people engage in all sorts of pleasant, diverting pursuits that carry risks, from rock climbing to beer drinking (don't mix those two), and if we banned all of them, we'd die of boredom.

What the evidence does show is that while video-gaming might make you a little better at certain jobs that demand visual acuity under stress, like piloting a fighter jet or performing an appendectomy, it's not going to make you generally smarter. And if you do a whole lot of it, it may well leave you scatterbrained—less able to concentrate on a single task, particularly a difficult one. More broadly, we should be skeptical of anyone who draws on video game studies to argue that spending a lot of time in front of a computer screen strengthens our attentiveness or our memory or even our ability to multitask. Taken as a whole, the evidence, including the video-gaming evidence, suggests it has the opposite effect.

MEMORY IS THE GRAVITY
OF MIND

July 14, 2011

> As gravity holds matter from flying off into space, so memory
> gives stability to knowledge; it is the cohesion which keeps
> things from falling into a lump, or flowing in waves.

> —*Ralph Waldo Emerson*

A FASCINATING AND UNSETTLING study of the internet's effects
on memory has just come out in *Science*. It provides more evidence of
how quickly our minds adapt to the tools we use to think with, for
better and for worse.

The study, "Google Effects on Memory: Cognitive Consequences of
Having Information at Our Fingertips," was conducted by three psychologists: Betsy Sparrow, of Columbia; Jenny Liu, of the University
of Wisconsin; and Daniel Wegner, of Harvard. They ran a series of
experiments aimed at answering this question: Does our awareness of
our ability to use Google to quickly find any fact or other bit of information influence the way our brains form memories? The answer, they
discovered, is yes: "When people expect to have future access to information, they have lower rates of recall of the information itself and
enhanced recall instead for where to access it." The findings suggest,
the researchers write, "that processes of human memory are adapting
to the advent of new computing and communication technology."

In the first experiment, people were asked a series of trivia questions. They were then given a test in which they were shown different
corporate brand names, some from search engines (e.g., Google) and
some from other familiar companies (e.g., Nike), in different colors

and asked to identify the color. In this kind of test, called a Stroop task, a greater delay in naming the color indicates a greater interest in, and cognitive focus on, the word itself. As the researchers explain: "People who have been disposed to think about a certain topic typically show slowed reaction times for naming the color of the word when the word itself is of interest and is more [cognitively] accessible, because the word captures attention and interferes with the fastest possible color naming." The experiment revealed that after people are asked a question to which they don't know the answer, they take significantly longer to identify the color of a search-related brand name than a non-search-related one. "It seems that when we are faced with a gap in our knowledge," the psychologists conclude, "we are primed to turn to the computer to rectify the situation." We've trained our brains to immediately think of using a computer when we're called on to answer a question or otherwise provide some bit of knowledge.

In the second experiment, people read forty factual statements of the kind you'd tend to look up with a search engine (e.g., "an ostrich's eye is bigger than its brain") and then typed the statements into a computer. Half the participants were told the computer would save what they typed, and half were told that what they typed would be erased. Afterward, the participants were asked to write down as many of the statements as they could remember. The experiment revealed that people who believed the information would be stored in the computer had a weaker memory of the information than those who assumed that the information would not be saved. "Participants apparently did not make the effort to remember when they thought they could later look up the trivia statements they had read. Since search engines are continually available to us, we may often be in a state of not feeling we need to encode the information internally. When we need it, we will look it up."

In the last experiment, people again read a series of factual statements and typed them into a computer. They were told that the statements would be stored in a specific folder with a generic name (e.g., "facts" or "data"). They were then given ten minutes to write down as many statements as they could remember. Finally, they were asked to name the folder in which a particular statement was stored. It was discovered that people were better able to remember the folder names than the facts themselves. "These results seem remarkable on the surface, given the memorable nature of the statements and the unmemorable nature of the folder names," the researchers write. The experiment suggests "that when people expect information to remain continuously available (such as we expect with Internet access), we are more likely to remember where to find it than we are to remember the details of the item."

Human beings have always had external—or, to use the psychological jargon, "transactive"—information stores to supplement their biological memory. These stores can reside in the brains of other people we know (if your friend John is an expert on sports, then you know you can use John's knowledge of sports facts to supplement your own memory) or in storage or media technologies such as maps and books. But we've never had an external memory store so capacious, so available, and so easily searched as the web. If, as this study indicates, the way we form (or fail to form) memories is deeply influenced by the mere existence of external information stores, then we may be entering an era in which we will store fewer and fewer memories inside our own brains.

If a fact stored in a computer were the same as a memory of that fact stored in the mind, the loss of internal memory wouldn't much matter. But external storage and biological memory are not the same thing. When we form, or "consolidate," a personal memory, we also form associations between that memory and other memories. The associations, unique to every mind, are indispensable to the development of deep, conceptual knowledge. And the associations are organic, not mechanical. They continue to change with time, as we learn more and experience more. As Emerson understood, the essence of personal memory is not the discrete facts or experiences we store in our mind but "the cohesion" that ties all those facts and experiences together.

The researchers seem fairly sanguine about the results of their study.

"We are becoming symbiotic with our computer tools," they conclude, "growing into interconnected systems that remember less by knowing information than by knowing where the information can be found." Although we don't yet understand the possible "disadvantages of being constantly 'wired,'" we have nevertheless "become dependent" on our gadgets. "We must remain plugged in to know what Google knows." But as memory shifts from the individual mind to the machine's shared database, what happens to that unique "cohesion" that gives rise to personal knowledge, selfhood's core?

THE MEDIUM IS McLUHAN

July 18, 2011

ONE OF MY FAVORITE YouTube videos is a clip from a 1968 Canadian TV show featuring a debate between Norman Mailer and Marshall McLuhan. The two men, both icons of the sixties, could hardly be more different. Leaning forward in his chair, Mailer is pugnacious, animated, engaged. McLuhan, abstracted and smiling wanly, seems to be on autopilot. He speaks in canned riddles. "The planet is no longer nature," he announces, to Mailer's uncomprehending stare; "it's now the content of an art work."

Watching McLuhan (who would have turned one hundred this week), you can't quite decide whether he was a genius or just had a screw loose. Both impressions, it turns out, are justified. As the novelist Douglas Coupland argued in his recent biography, *Marshall McLuhan: You Know Nothing of My Work!*, McLuhan's mind was probably situated at the mild end of the autism spectrum. He also suffered from a couple of major cerebral traumas. In 1960, he had a stroke so severe that he was given his last rites. In 1967, just a few months before the Mailer debate, surgeons removed a tumor the size of a small apple from the base of his brain. A later procedure revealed that he had an extra artery pumping blood into his cranium.

Between the stroke and the tumor, McLuhan managed to write a pair of extravagantly original books. *The Gutenberg Galaxy*, published in 1962, explored the cultural and personal consequences of the printing press, arguing that Gutenberg's invention shaped the modern mind. Two years later, *Understanding Media* extended the analysis to the electric media of the twentieth century, which, McLuhan argued, were destroying the individualist ethos of print culture and turning the world into a tightly networked global village. The ideas in both books drew heavily on the works of other thinkers, including such contemporaries as Harold Innis, Albert Lord, and Wyndham Lewis, but McLuhan's synthesis was, in content and tone, unlike anything that had come before.

When you read McLuhan today, you find all sorts of reasons to be impressed by his insight into media's far-reaching effects and by his anticipation of the course of technological progress. When he looked at a Xerox machine in 1966, he didn't just see the ramifications of cheap photocopying, as great as they were. He foresaw the transformation of the book from a manufactured object into an information service: "Instead of the book as a fixed package of repeatable and uniform character suited to the market with pricing, the book is increasingly taking on the character of a service, an information service, and the book as an information service is tailor-made and custom-built." That must have sounded outrageous a half-century ago. Today, with books shedding their physical skins and turning into software programs, it sounds like a given.

You also realize that McLuhan got a whole lot wrong. One of his central assumptions was that electric communication technologies would displace the phonetic alphabet from the center of culture, a process he felt was well under way in his own time. "Our Western values, built on the written word, have already been considerably affected by the electric media of telephone, radio, and TV," he wrote in *Understanding Media*. He believed that readers, because their attention is consumed by the act of interpreting the visual symbols of alphabetic letters, become alienated from their other senses, sacrifice their attachment to other people, and enter a world of abstraction, individualism, and rigorously linear thinking. This, for McLuhan, was the story of Western civilization, particularly after the arrival of Gutenberg's press.

By freeing us from our single-minded focus on the written word, new technologies like the telephone and the television would, he argued, broaden our sensory and emotional engagement with the world and with others. We would become more integrated, more "holistic," at both a personal and a social level, and we would recoup some of our primal nature. But McLuhan failed to anticipate that, as the speed and capacity of communication networks grew, what they would end up transmitting more than anything else is text. The written word would invade electric media. If McLuhan were to come back to life today, the sight of people using their telephones as reading and writing devices would blow his mind. He would also be amazed to discover that the fuzzy, low-definition TV screens that he knew (and on which he based his famous distinction between hot and cool media) have been replaced by crystal-clear, high-definition monitors, which more often than not are crawling with the letters of the alphabet. Our senses are more dominated by the need to maintain a strong, narrow visual focus than ever before. Electric media are social media, but they are also media of isolation. If the medium is the message, then the message of electric media has turned out to be far different from what McLuhan supposed.

That some of his ideas didn't pan out wouldn't have bothered him much. He was far more interested in playing with ideas than nailing them down. He intended his writings to be "probes" into the present and the future. He wanted his words to knock readers out of their intellectual comfort zones, to get them to entertain the possibility that their accepted patterns of perception might need reordering. Fortunately for him, he arrived on the scene at a rare moment in history when large numbers of people wanted nothing more than to have their minds messed with.

McLuhan was a scholar of literature, with a doctorate from Cambridge, and his interpretation of the intellectual and social effects of media was richly allusive and erudite. But what particularly galvanized the public and the press was the weirdness of his prose. Perhaps as a consequence of his unusual mind, he had a knack for writing sentences that sound at once clinical and mystical. His books read like accounts of acid trips written by a bureaucrat. That kaleidoscopic, almost psychedelic style made him a darling of the counterculture—the bearded and the Birkenstocked embraced him as a guru—but it alienated him from his colleagues in academia. To them, McLuhan was a celebrity-seeking charlatan.

Neither his fans nor his foes saw him clearly. The central fact of McLuhan's life was his conversion, at the age of twenty-five, to Catholicism, and his subsequent devotion to the religion's rituals and tenets. He became a daily Mass-goer. Though he never discussed it, his faith forms the moral and intellectual backdrop to all his mature work. What lay in store, McLuhan believed, was the timelessness of eternity. The earthly conceptions of past, present, and future were by comparison of little consequence. His role as a thinker was not to celebrate or denigrate the world but simply to understand it, to recognize the patterns that would unlock history's secrets and thus provide hints of God's design. His job was not dissimilar, as he saw it, from that of the artist.

That's not to say that McLuhan was without secular ambition. Coming of age at the dawn of mass media, he very much wanted to be famous. "I have no affection for the world," he wrote to his brother in the late thirties, at the start of his academic career. But in the same letter he disclosed the "large dreams" he harbored for "the bedazzlement of men." Modern media needed its own medium, the voice that would explain its transformative power to the world, and he would take that role.

The tension between McLuhan's craving for earthly attention and his distaste for the material world would never be resolved. Even as he came to be worshipped as a techno-utopian seer in the mid-sixties, he had already, writes Coupland, lost all hope "that the world might become a better place with new technology." He heralded the global village, and was genuinely excited by its imminence and its possibilities, but he also saw its arrival as the death knell for the literary culture he revered. The electronically connected society would be the setting not for the further flourishing of civilization but for the return of tribalism, if on a vast new scale. "And as our senses [go] outside us," he wrote, "Big Brother goes inside." Always on display, always broadcasting, always watched, we would become mediated, technologically and socially, as never before. The intellectual detachment that characterizes the solitary thinker—and that was the hallmark of McLuhan's own work—would be replaced by the communal excitements, and constraints, of what we have today come to call "interactivity."

McLuhan also saw, with biting clarity, how all mass media are fated to become tools of commercialism and consumerism—and hence instruments of control. The more intimately we weave media into our lives, the more tightly we become locked in a corporate embrace. As he wrote in *Understanding Media*, "Once we have surrendered our senses and nervous systems to the private manipulation of those who would try to benefit by taking a lease on our eyes and ears and nerves, we don't really have any rights left." Has a darker vision of modern media ever been expressed?

"Many people seem to think that if you talk about something recent, you're in favor of it," McLuhan explained during an uncharacteristically candid interview in 1966. "The exact opposite is true in my case. Anything I talk about is almost certain to be something I'm resolutely against, and it seems to me the best way of opposing it is to understand it, and then you know where to turn off the button." Though the founders of *Wired* magazine would posthumously appoint McLuhan the "patron saint" of the digital revolution, the real McLuhan was as much a Luddite as a technophile. He would have found the collective banality of Facebook abhorrent, if also fascinating.

In the fall of 1979, McLuhan suffered another major stroke, but this was one from which he would not recover. Though he regained consciousness, he remained unable to read, write, or speak until his death a little more than a year later. A lover of words—his favorite book was Joyce's *Finnegans Wake*—he died in a state of wordlessness. He had fulfilled his own prophecy and become postliterary.

FACEBOOK'S BUSINESS MODEL

September 26, 2011

THE DESIRE FOR PRIVACY is strong; vanity is stronger.

UTOPIA IS CREEPY

October 29, 2011

WORKS OF SCIENCE FICTION, at least the good ones, are almost always dystopian. It's easy to see why. There's a lot of drama in hell, but heaven is, by definition, conflict-free. Happiness may be nice to experience, but seen from the outside it's pretty dull.

There's another reason portrayals of utopia don't work. We've all experienced the "uncanny valley" that makes it difficult to watch robotic replicas of human beings without being creeped out. The uncanny valley also exists when it comes to viewing artistic renderings of a future paradise. Utopia is creepy—or at least it looks creepy. That's probably because utopia requires its residents to behave like robots, never displaying or even feeling fear or anger or jealousy or despair or bitterness or any of those other messy emotions that plague our fallen world.

I've noticed the arrival recently of a new genre of futuristic YouTube video. The films are created by technology companies for marketing or brand-burnishing purposes. With the flawless production values that only a cash-engorged balance sheet can buy you, they portray a not-too-distant future populated by exceedingly well-groomed people who spend their hyperproductive days going from one computer display to the next. (As seems always to be the case with utopias, the atmosphere is very postsexual.)

The latest comes from Microsoft—it bears the evocative title "Productivity Future Vision (2011)"—and like its predecessors it seems to be the product of a collaboration between Stanley Kubrick and David Lynch. It opens with a sleek, black-clad businesswoman (human, I think) walking through an airport after a flight. She touches her computerized eyeglasses and a digitized voice directs her outside, where a section of the street lights up to define a personal "pickup zone." Her limo arrives promptly, and as soon as she settles into the backseat the car's windows turn into computer monitors, displaying her upcoming schedule and streaming other practical information. Her phone, meanwhile, transmits her estimated time of arrival to a hotel bellhop, who tracks her approach through a screen the size of a business card. The video then cuts away from the businesswoman and proceeds through a series of other, equally sterile vignettes. Lots of manicured fingers swipe lots of highly responsive panes of glass. Arrays of data float through the air. Deals get done hologrammatically. The film ends with a family silently inhabiting a kitchen where every surface is a graphical user interface. The people look small, insufficient.

Designed to get us jazzed about the future, these kinds of productions strive to conjure up visions of technological Edens. But they end up doing the opposite: portraying a future that feels cold, mechanical, and repellent. Accentuating the creepiness are the similarities between the world they project and the one we live in.

SPINELESSNESS

March 16, 2012

THE PRINT EDITION OF the Encyclopedia Britannica has been consigned to history's dumpster. The announcement, hardly unexpected, came two days ago: From now on, the venerable, multivolume reference work will be available only in digital form. The weighty shall be weightless. It's the way of the world.

I'm going to miss the spines—all forty-five of them, ranked across the shelf like stoic beefeaters. They're handsome things, somehow managing to be imposing and inviting at the same time. The best part is that each one is branded with a pair of index words, there to tell you where the volume begins and where it ends. Some of these near-random, two-word phrases don't rise above mere functionality: **India Ireland**, for instance, or **Accounting Architecture.** Others, though, open up new and unexpected territory for the imagination to wander in.

Here, for the record, are a few of my favorites:

Freon Holderlin (a man I'd like to meet, despite his reputation for coldness)

Menage Ottawa (a perfect oxymoron)

Chicago Death (Jack White's new side project)

Light Metabolism (what the Theory of Everything, once discovered, will be called)

Excretion Geometry (a field understood by only seven people in the world, all now deceased)

Arctic Biosphere (Freon Holderlin lives here, according to rumor)

Krasnokamsk Menadra (when I take up meditation, this will be what I chant)

And—I'm starting to choke up—my favorite one of all:

Decorative Edison

FUTURE GOTHIC

May 8, 2012

1.

Hardware is a problem. It wears out. It breaks down. It is subject to physical forces. It is subject to entropy. It deteriorates. It decays. It fails. The moment of failure can't be predicted, but what can be predicted is that the moment will come. Assemblages of atoms are doomed. Worse yet, the more components incorporated into a physical system—the more subassemblies that make up the assembly—the more points of failure the apparatus has and the more fragile it becomes.

This is an engineering problem. This is also a metaphysical problem.

2.

One of Google's great innovations in building the data centers that run its searches was to use software as a means of isolating each component of the system and hence of separating component failure from system failure. The networking software senses a component failure (a dying hard drive, say) and immediately bypasses the component, routing the work to another, healthy piece of hardware in the system. No single component matters; each is dispensable and disposable. Maintaining the system, at the hardware level, becomes a simple process of replacing failed parts with fresh ones. You hire a low-skilled worker, or build a robot, and when a component dies, the worker, or the robot, swaps it out with a good one.

Such a system requires smart software. It also requires cheap parts.

3.

Executing an algorithm with a physical system is like putting a mind into a body.

4.

Bruce Sterling, the cyberpunk writer, gave an interesting speech at a European tech conference a couple years back. He drew a distinction between two lifestyles that form the poles of our emerging culture. At one end—the top end—you have "Gothic High-Tech."

> In Gothic High-Tech, you're Steve Jobs. You've built an iPhone, which is a brilliant technical innovation, but you also had to sneak off to Tennessee to get a liver transplant because you're dying of something secret and horrible. And you're a captain of American industry. You're not some General Motors kinda guy. On the contrary, you're a guy who's got both hands on the steering wheel of a functional car. But you're still Gothic High-Tech because death is waiting. And not a kindly death either, but a sinister, creeping, tainted wells of Silicon Valley kind of Superfund thing that steals upon you month by month, and that you have to hide from the public and from the bloggers and from the shareholders.

And then there's the other end—the bottom end—which Sterling calls "Favela Chic." It comprises the multitudinous "play-laborers" of the virtual realm.

> Favela Chic is when you have lost everything material, everything you built and everything you had, but you're still wired to the gills! And really big on Facebook. That's Favela Chic. You lost everything, you have no money, you have no career, you have no health insurance, you're not even sure where you live, you don't have chil-

dren, and you have no steady relationship or any set of dependable friends. And it's hot. It's a really cool place to be.

The Favela Chic worship the Gothic High-Tech because the Gothic High-Tech have perfected unreality. They have escaped the decaying realm of "the infrastructure" and have positioned themselves "in the narrative," the stream that flows forever, unimpeded. They are avatars: software without apparatus, mind without body.

Except when a part fails.

5.

H. G. Wells, in his 1895 Gothic novella *The Time Machine*, used different terms. The Gothic High-Tech he called Morlocks. The Favela Chic he called Eloi. Of course Wells was writing in a time of industrialization rather than virtualization.

About eight or nine in the morning I came to the same seat of yellow metal from which I had viewed the world upon the evening of my arrival. I thought of my hasty conclusions upon that evening and could not refrain from laughing bitterly at my confidence. Here was the same beautiful scene, the same abundant foliage, the same splendid palaces and magnificent ruins, the same silver river running between its fertile banks. The gay robes of the beautiful people moved hither and thither among the trees. Some were bathing in exactly the place where I had saved Weena, and that suddenly gave me a keen stab of pain. And like blots upon the landscape rose the cupolas above the ways to the Under-world. I understood now what all the beauty of the Over-world people covered. Very pleasant was their day, as pleasant as the day of the cattle in the field. Like the cattle, they knew of no enemies and provided against no needs. And their end was the same.

6.

The young, multibillionaire technologist is left with only two avocations: space travel and the engineering of immortality. Both are about escaping the gravity of the situation.

There are a couple of ways to sidestep death. You can virtualize the apparatus, freeing the mind from the body. But before you can do that, you need to figure out the code. And, alas, when it comes to human beings, we are still a long way from figuring out the code. Disembodiment is not imminent. Or you can take the Google route and figure out a way to bypass any failed component, whether it's heart or kidney, pancreas or liver. In time, we may figure out a way to fabricate the essential components of our bodies—to create an unlimited supply of parts—but that eventuality, too, is not imminent. So we are left, for the time being, with transplantation, with the harvesting of good components from failed systems and the use of those components to replace the failed components of living systems.

The Gothic High-Tech, who cannot abide death, face a problem here: The organ donation system is largely democratic; it can't be gamed easily by wealth. A rich person may be able to travel somewhere that has shorter lines—Tennessee, say—but he can't jump to the head of the line. So the challenge becomes one of increasing the supply, of making rare components plentiful.

7.

A week ago, Facebook CEO Mark Zuckerberg, in a move that he said was inspired by the experience of his friend Steve Jobs, announced that Facebook was introducing a new feature that would make it easy for members to identify themselves as organ donors. Should Zuckerberg's move increase the supply of organs, it will save many lives and alleviate much suffering. We should all be grateful. Dark dreams of the future are best left to science fiction writers.

THE HIERARCHY OF INNOVATION

May 14, 2012

> If you could choose only one of the following two inventions,
> indoor plumbing or the Internet, which would you choose?
>
> —*Robert J. Gordon*

HARVARD BUSINESS REVIEW EDITOR Justin Fox is the latest pundit to ring the "innovation ain't what it used to be" bell. "Compared with the staggering changes in everyday life in the first half of the 20th century," he writes, "the digital age has brought relatively minor alterations to how we live." Fox has a lot of company. He points to science fiction writer Neal Stephenson, who worries that the internet, far from spurring a great burst of industrial creativity, may have put innovation "on hold for a generation." Fox also cites economist Tyler Cowen, who has argued that, recent techno-enthusiasm aside, we're living in a time of innovation stagnation.

He might also have mentioned tech powerbroker Peter Thiel, who believes that large-scale innovation has gone dormant and that we've entered a technological "desert." Thiel blames the hippies. "Men reached the moon in July 1969," he wrote in "The End of the Future," a 2011 *National Review* article, "and Woodstock began three weeks later. With the benefit of hindsight, we can see that this was when the hippies took over the country, and when the true cultural war over Progress was lost."

The original inspiration for such grousing—about progress, not hippies—is a Northwestern University economist named Robert J. Gordon, whose 2000 paper "Does the 'New Economy' Measure Up to the

Great Inventions of the Past?" included a damning comparison of the flood of inventions that occurred a century ago with the seeming trickle that we see today. Consider just a few of the products invented in the ten years between 1876 and 1886: internal combustion engine, electric lightbulb, electric transformer, steam turbine, electric railroad, automobile, telephone, movie camera, phonograph, linotype, cash register, vaccine, reinforced concrete, flush toilet. The typewriter had arrived a few years earlier and the punch-card tabulator (the forerunner of the digital computer) would appear a few years later. Then, in short order, came airplanes, radio, air-conditioning, the vacuum tube, jet aircraft, television, refrigerators and a raft of other home appliances, as well as revolutionary advances in manufacturing processes. (And let's not forget The Bomb.) The conditions of life changed utterly between 1890 and 1950, observed Gordon. Between 1950 and today? Not so much.

So why is innovation less impressive today? Maybe Thiel is right, and it's the fault of hippies, liberals, and other degenerates. Or maybe it's crappy education. Or a lack of corporate investment in research. Or shortsighted venture capitalists. Or overaggressive lawyers. Or imagination-challenged entrepreneurs. Or maybe it's a catastrophic loss of American mojo. None of these explanations makes much sense, though. The aperture of science grows ever wider, after all, even as the commercial and reputational rewards for innovation grow, investment pools swell, and the ability to share ideas strengthens. Any barrier to innovation should be swept away by such forces.

Let me float an alternative explanation: There has been no decline in innovation; there has just been a shift in its focus. We're as creative as ever, but we've funneled our creativity into areas that produce smaller-scale, less far-reaching, less visible breakthroughs. And we've done that for entirely rational reasons. We're getting precisely the kind of innovation that we desire—and that we deserve.

My idea is that there's a hierarchy of innovation that runs in parallel with Abraham Maslow's famous hierarchy of needs. Maslow argued that human needs progress through five stages, with each new stage requiring the fulfillment of lower-level, or more basic, needs. So first we need to meet our most primitive physiological needs, and that frees us to focus on our needs for safety, and once our needs for safety are met, we can attend to our needs for belongingness, and then on to our needs for personal esteem, and finally to our needs for self-actualization. If you look at Maslow's hierarchy as an inflexible structure, with clear boundaries between its levels, it falls apart. Our needs are messy, and the boundaries between them are porous. A caveman probably pursued self-esteem and self-actualization, to some degree, just as we today expend effort seeking to fulfill our physical needs. But if you look at the hierarchy as a map of human focus, or emphasis, then it makes sense—and indeed seems to be borne out by history. In short: The more comfortable you are, the more time you spend thinking about yourself.

If inventors respond to human needs (as they surely do, particularly when acting for commercial gain), then shifts in needs would necessarily bring shifts in the focus of technological innovation and material progress. The tools we invent would move up through the hierarchy of needs, from tools that help us safeguard our bodies to tools that allow us to modify our internal states, from tools of survival to tools of the self. Here's my crack at what the hierarchy of innovation looks like:

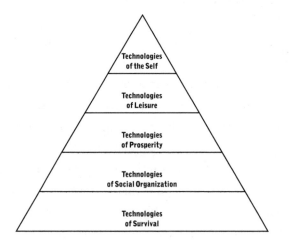

The Hierarchy of Innovation

History suggests that, at least in the economically developed world, the focus of innovation has indeed moved upward through these five stages, propelled by shifts in the needs people seek to fulfill. In the beginning came technologies of survival. Our overriding need was to stay alive, and so we invented rudimentary shelters, weapons for hunting and self-defense, cooking implements, and articles of clothing. Once we felt reasonably secure, we began to form stable societies. That required a broad set of technologies of social organization. We erected castles and churches, made farming tools and weapons of war, built aqueducts and latrines. Once we had working social organizations, and the division of labor that went with them, we turned our attention to technologies of prosperity, and we started creating complex systems of transportation, education, medicine, energy, manufacturing, trade, and communication. Having achieved some prosperity, and gained some disposable income, we began to crave technologies of leisure: consumer goods and services, entertainments, tourism, radios and TVs, stylish cars and clothes. Now, having accumulated enough stuff to fulfill our material needs, our desires are turning to self-actualization and self-expression. What we now want are technologies of the self. Think Xanax and Viagra. Think cosmetic surgery and stem-cell-infused antiaging creams. Think Facebook and Twitter and Pinterest. Think Fitbit.

As with Maslow's hierarchy of needs, the hierarchy of innovation is not a rigid one. Innovation today continues at all five levels. But the rewards, both monetary and reputational, are greatest for inventors and entrepreneurs who focus on technologies of the self, and that's why so much investment and entrepreneurial activity has moved in that direction. We're already physically comfortable, so getting a little more physically comfortable doesn't seem particularly pressing. We've become inward looking, and what we desire are more powerful tools for modifying our internal state (to gain happiness, or at least a pleasant numbness) or projecting that state outward (to gain status, or at least attention). A start-up has a greater prospect of renown and riches if it creates, say, a popular social networking app than if it launches an effort to build a faster, more efficient system for mass transit.

As we move toward the top level of the innovation hierarchy, our inventions have less visible, less transformative effects. We're no longer changing the shape of the physical world or even the shape of society as it manifests itself in the physical world. We're altering moods and identities, reshaping the invisible self. When you step back and take a broad view, it looks like stagnation—it looks like nothing much is going on. When you contrast what's happening today with what happened a hundred years ago, when our focus on technologies of prosperity was peaking and our focus on technologies of leisure was sharpening, it's hard not to see an innovation desert.

The pursuit of self-actualization, Maslow argued, leads ultimately to a healthy and whole individual. It brings "a fuller knowledge of, and acceptance of, the person's own intrinsic nature, as an unceasing trend toward unity, integration or synergy within the person." When reduced to the purchase and use of technological goods, however, the pursuit of self-actualization turns quickly into mere self-indulgence. Innovation, to put a dark spin on it, arcs toward decadence. But we shouldn't be too harsh in judging our innovators. They're just giving us what we want.

RIP. MIX. BURN. READ.

June 4, 2012

> So I don't buy records in your shop,
> Now I tape them all 'cause I'm Top of the Pops.
>
> —*Bow Wow Wow, "C30 C60 C90 Go!," 1980*

WHEN I TURNED TWELVE, in the early seventies, I received, as a birthday present from my parents, a portable, Realistic-brand cassette tape recorder from Radio Shack. Within hours, I became a music pirate. I had a friend who lived next door, and his older brother had a copy of *Abbey Road*, an album I had taken a shine to. I carried my recorder over to their house, set its little plastic microphone (it was a mono machine) in front of one of the speakers of their family stereo, and proceeded to make a cassette copy of the record. I used the same technique at my own house to record hits off the radio and make copies of my siblings' and friends' LPs and 45s. It never crossed my mind that I was doing anything wrong. I didn't think of myself as a pirate, and I didn't think of my recordings as being illicit. I was just being a fan.

I was not alone. Tape recorders, whether reel-to-reel or cassette, were everywhere, and pretty much any kid who had access to one made copies of albums and songs. (If you've read Walter Isaacson's biography of Steve Jobs, you know that when Jobs went off to college in 1972, he brought with him a comprehensive collection of Dylan bootlegs on tape.) When, a couple of years later, cassette decks became commonplace components of stereo systems, ripping songs from records and the radio became even simpler. There was a reason that cassette decks had input jacks as well as output jacks. My friends and I routinely exchanged albums and mixtapes on cassettes. It was the norm.

We also, I should point out, bought a lot of records, particularly when we realized that pretty much everything being played on the radio was garbage. (I apologize to all Doobie Brothers fans who happen to be reading this.) There are a few reasons why record sales and record copying flourished simultaneously. First, in order to make a copy of an album, someone in your circle of friends had to own an original; there were no anonymous, long-distance exchanges of tunes. Second, vinyl was a superior medium to tape because, among other things, it made it easier to play individual tracks (and it was not unusual to play a favorite track over and over again). Third, record jackets were cool and they had considerable value in and of themselves. Fourth, owning the record had social cachet. And fifth, records weren't that expensive. What a lot of people forget about LPs back then is that most of them, not long after their original release, were remaindered as what were called cutouts, and you could pick them up for $1.99 or so. Even as a high-schooler working a part-time, minimum-wage job, you could afford to buy a record a week, which was—believe it or not—plenty.

The reason I'm telling you all this is not that I suddenly feel guilty about my life as a teenage music pirate. I feel no guilt whatsoever. It's just that this weekend I happened to read an article in the *Wall Street Journal* by Listen.com founder Rob Reid, who argued that "in the swashbuckling arena of digital piracy, the publishing world is acquitting itself far better than the brash music industry." Drawing a parallel between the music and book businesses, Reid observed that publishing "is now further into its own digital history than music was when Napster died."

Both histories began when digital media became portable. For music, that was 1999, when the record labels ended a failing legal campaign to ban MP3 players. For books, it came with the 2007 launch of the Kindle. Publishing has gotten off to a much better start. Both industries saw a roughly 20% drop in physical sales four years after their respective digital kickoffs. But ebook sales have largely made up the shortfall in publishing—unlike digital music sales, which stayed stubbornly close to zero for years.

This doesn't prove that music lovers are crooks. Rather, it shows that actually selling things to early adopters is wise. Publishers did this—unlike the record labels, which essentially insisted that the first digital generation either steal online music or do without it entirely.

That all seems sensible enough. But Reid's argument is misleading. He oversimplifies media history, and he glosses over some big and fundamental differences between the book market and the music market. As my own youthful experience suggests, music lovers *are* crooks, and we've been crooks for decades. ("Crooks" is his term, of course, not mine.) Moreover, the "digital history" of music did not begin in 1999. It began in 1982 when albums began to be released on compact disk. Yes, there are some similarities between the music and book industries, and they're worth attending to, but the fact that the two industries have taken different courses in the digital era probably has far more to do with the basic differences between them—differences in history, technology, and customers, among other things—than with differences in executive decision making.

Let me review some of the most salient differences and the way they've influenced the divergent paths the industries have taken:

Kids copied music long before music went digital. The unauthorized copying of songs and albums did not begin with the arrival of the web or of MP3s or of Napster. It has been a part of the culture of pop music since the sixties. There has been no such tradition with books. Xeroxing a book was not an easy task, and it was fairly expensive, to boot. Nobody did it, except, maybe, the occasional oddball. So, even though the large-scale trading of bootlegged songs made possible by the net had radically different implications for the music business than the small-scale trading that had taken place previously, digital copying and trading didn't feel particularly different from making and exchanging tapes. It seemed like a new variation on an old practice.

Fidelity matters less for popular music than for books. This seems counterintuitive, but it's true. I was happy with my copy of *Abbey Road* despite its abysmal sound quality and the fact that—cringe!—I had only recorded one channel of a stereo mix. Throughout the sixties and well into the seventies, the main way a lot of people listened to music was through crappy car radios and crappy transistor radios. (Need I mention eight-tracks?) The human ear and the human brain seem to be adept at turning lo-fi music signals into fulfilling listening experiences—the auditory imagination somehow fills in the missing signal. Early MP3s, though they were often ripped at very low bit rates, sounded just fine to the vast majority of the music-listening public, so quality was no barrier to mass piracy. A lo-fi copy of a book, in contrast, is a misery to read. Blurry text, missing pages, clunky navigation: it takes a very dedicated reader to overcome even fairly minor shortcomings in a copy of a book. That's one of the main reasons that even though bootlegged copies of popular books have been freely available online for quite some time now, in the form of scans, few people bother with them.

Books never had a CD phase. Music was digitized long before the arrival of the web. During the eighties, record companies digitized their catalogs, and digital CDs soon displaced tapes and vinyl as the medium of choice for music. The transition was a boon to the music industry because a whole lot of consumers bought new CD copies of albums that they already owned on vinyl. But the boon (as Reid notes) also set the stage for the subsequent bust. When personal computers with CD-ROM drives made it possible to rip music CDs into MP3 files, all the music that most people would ever want was soon available in a form that could be easily exchanged online. The CD also had the unintended effect of making the physical record album less valuable. CD cases were small, plastic, and annoying; the booklets wedged inside them were rarely removed; and the disks themselves had a space-age sterility that rendered them entirely charmless. By reducing the perceived value of the physical product, CDs made it easier for consumers to discard that product. The book business did not go through a digitization phase prior to the arrival of the web, so there was no supply of digital books waiting to be traded when online trading became possible. It was an entirely different situation from a technological standpoint.

The average music buyer is younger than the average book buyer. Young people have long been the primary market for popular music. Young people also tend to have the spare time, the tech savvy, the obliviousness to risk, the constrained wallets, and the passion for music required to do a whole lot of bootlegging. Books tend to be sold to older people. Older people make lousy pirates. They're busy, they have money, and they just don't care that much. That's another crucial reason why book publishers have been sheltered from piracy in a way that record companies weren't.

When Apple first promoted its iTunes app—this was quite a while before it got into music retailing—it used the slogan "Rip. Mix. Burn." Though it wouldn't admit it, Apple wanted people to engage in widespread copying and trading of music, because the more free digital music files that went into circulation, the more attractive its computers (and subsequently its iPods) became. That slogan never had an analog in the book business because the history, technology, and customers of the book business were very different at the start of the digital age. People like Reid like to suggest that if record company executives had made different decisions a decade ago, the fate of their industry would have been different. I'm skeptical about that. Sure, they could have made different decisions, but I really don't think it would have changed the course of history much. They were basically screwed.

And executives in the publishing industry are probably kidding themselves if they think that they're responsible for the fact that, at least so far, their business hasn't gone through the wrenching changes that have affected their peers in the music business. They were just lucky to be peddling books instead of songs.

LIVE FAST, DIE YOUNG, AND LEAVE A BEAUTIFUL HOLOGRAM

June 12, 2012

"FOR US, OF COURSE, it's about keeping Jimi authentically correct." That's Janie Hendrix, explaining the motivation behind her effort to turn her long-dead brother into a Strat-wielding hologram. Tupac Shakur's recent leap from grave to stage at the Coachella festival was just the opening act in what promises to be an orgy of cultural necrophilia. *Billboard* reports that hologrammatic second comings are in the works not just for Jimi Hendrix but for Elvis Presley, Jim Morrison, Otis Redding, Janis Joplin, Peter Tosh, and even Rick James. Superfreaky! What could be more authentically correct than an image of an image?

I'm really looking forward to seeing the Doors with Jim Morrison back out in front—that guy from the Cult never did it for me—but I admit it may be a little unsettling to see the rest of the band looking semi-elderly while the Lizard King appears as his perfect, leather-clad, twenty-four-year-old self. Jeff Jampol, the Doors' manager, says, "Hopefully, 'Jim Morrison' will be able to walk right up to you, look you in the eye, sing right at you and then turn around and walk away." That's all well and good, but I'm sure Jampol knows that the crowd isn't going to be satisfied unless "Jim Morrison" whips out his virtual willy. (Can you arrest a hologram for obscenity?) In any event, hearing the Morrison hologram sing "Cancel my subscription to the resurrection" is going to be priceless—a once-in-a-lifetime moment, replayable endlessly.

I think it was Nietzsche who said that what kills you only makes you stronger in the marketplace.

ONLINE, OFFLINE, AND THE LINE BETWEEN

July 2, 2012

In "The IRL Fetish," an article in *The New Inquiry* that has tweeters atwitter, Nathan Jurgenson upbraids those who would draw a distinction between the real world and the virtual world, between "the offline" and "the online." Jurgenson, a sociology grad student, prefaces his screed by acknowledging the deepening "intrusion" of digital media into our everyday lives:

> Hanging out with friends and family increasingly means also hanging out with their technology. While eating, defecating, or resting in our beds, we are rubbing on our glowing rectangles, seemingly lost within the infostream.

Where's the Lysol?

But it's not just that we're spending so much time fondling our gizmos, Jurgenson says. It's that "the logic" of social networks and other online sites and services "has burrowed far into our consciousness." Computer software and related media shape not only our lives but our beings—our experiences, our perceptions, our relations with others. Back in the days of modem-equipped desktop computers, cyberspace was a well-demarcated place that you'd go and visit now and then. You'd log on, browse a while, then log off. It was easy to draw a line between online and offline. That's no longer the case. We're always connected. Online saturates offline.

So far, so good. Today we vibrate, sometimes pleasantly, sometimes uncomfortably, often unconsciously, between the virtual and the real. But then Jurgenson's argument goes topsy-turvy. Dismissing concerns that the net's omnipresence may be weakening our "real connection" to the world and one another, he suggests that the ubiquity of digital media is actually enriching our ties to the physical realm:

> We have never appreciated a solitary stroll, a camping trip, a face-to-face chat with friends, or even our boredom better than we do now. Nothing has contributed more to our collective appreciation for being logged off and technologically disconnected than the very technologies of connection. The ease of digital distraction has made us appreciate solitude with a new intensity. . . . In short, we've never cherished being alone, valued introspection, and treasured information disconnection more than we do now. Never has being disconnected—even if for just a moment—felt so profound.

You might say that Jurgenson is just stating the obvious, reprising the old Joni Mitchell refrain: "You don't know what you've got till it's gone." A really thirsty man will appreciate a glass of water more than an amply hydrated man. But instead of arriving at the obvious conclusion—that being amply hydrated is better than being really thirsty—Jurgenson gives the argument a wrench. The sense of loss that comes with being hyperconnected, he wants us to believe, is actually a sign of gain. "Nothing has contributed more to our collective appreciation for being logged off and technologically disconnected than the very technologies of connection." That sip of water was amazing! Thank God I'm parched! I guess you can't blame a guy for looking at the bright side, but while it's true that being without a precious thing makes that precious thing seem all the more precious, that hardly means we should celebrate the absence. A yearning for something slipping from our grasp should probably be taken as a warning.

But there are deeper problems with Jurgenson's assessment. What are we to make of this: "We have never appreciated a solitary stroll, a camping trip, a face-to-face chat with friends, or even our boredom better than we do now." That's the kind of sweeping statement that would benefit from a little evidence. A glance at the history of philosophy, literature, art, and recreation suggests that plenty of folks in the days before computer networks had a deep—a very, very deep—appreciation of the beauties and restorative qualities of solitude, nature, and long chats with friends. Our present age may be tops in some things, but it's far from tops in the area of solitary strolls.

The tragedy is that the stroll, the camping trip, and the face-to-face chat are now themselves suffused with digital ephemera. *Here* is under constant bombardment from *elsewhere*. Even if we agree to turn off our gadgets for a spell, they remain ghostly presences—all those missed messages hang like apparitions in the air, reminding us of our disconnectedness—and that serves to distance us from the unmediated experiences we seek.

Jurgenson doesn't even want to grant us license to acknowledge what's been lost. When someone expresses a yearning for direct experience of "the real world," he says, she's just "fetishizing the offline." She's indulging a reverence for something that, apparently, never really existed. "It is the fetish objects of the offline and the disconnected that are not real," he concludes, his argument becoming a tangle of abstractions. "Those who mourn the loss of the offline are blind to its prominence online." No, actually, they're not. The reason people struggle with the tension between online experience and offline experience is because there *is* a tension between online experience and offline experience, and people are smart enough to understand, to *feel*, that the tension does not evaporate as the online colonizes the offline. One need not subscribe to what Jurgenson calls "digital dualism"—"the habit of viewing the online and offline as largely distinct"—to believe that we sacrifice something real and important when we enter an environment crowded with computer screens and inundated with data transmissions.

Yes, the online is as much a part of "real life" as the offline—civilization and its inhabitants have always been, to borrow Walter Ong's term, "technologized"; reality has always been mediated—but the fact that the two realms of experience, the two states of being, are blurring, and blurring quickly, should spur us to think critically about the consequences of that blurring, not to conclude that the blurring turns a real distinction into a fiction, as if when you whisk oil and vinegar into a salad dressing, you whisk oil and vinegar out of existence. To exaggerate a distinction seems a lesser crime than to pretend it doesn't exist.

GOOGLE GLASS AND CLAUDE GLASS

September 19, 2012

GOOGLE COFOUNDER SERGEY BRIN made a stir earlier this month when he catted about New York Fashion Week with a Google Glass wrapped around his bean. It was something of a coming-out party for Google's reality-augmentation device, which promises to democratize the head-up display, giving us all a fighter pilot's view of the world. Diane von Furstenberg got Glassed. So did Sarah Jessica Parker. Wendi Murdoch seemed impressed by the cyborgian adornment, as did her husband, Rupert, who promptly tweeted, "Genius!" Google Glass is shaping up to be the biggest thing to hit the human brow since Olivia Newton-John's headband. Let's get post-physical.

It's apt that models, clothes designers, and others in the couture business would be among the first to embrace Glass. The fashion trade has always been at the forefront of reality augmentation. But Google's ocular accessory is not the first Glassware to come into vogue. In the eighteenth century, the gadget of choice for trendsetters was the Claude Glass. Named after the popular French landscape painter Claude Lorraine, the Claude Glass was a tinted, convex hand mirror that ladies and gentlemen would carry around on outings and whip out whenever they wanted to amp up the beauty of a natural scene. As Leo Marx explained in *The Machine in the Garden*, "When a viewer used the Claude Glass the landscape was transformed into a provisional work of art, framed and suffused by a golden tone like that of the master's paintings." The glass "helped create a pastoral illusion."

Where a Claude Glass bathed landscapes in a soft painterly light, a Google Glass bathes them in hard data. It gives its owner the eyes not of an artist but of an analyst. Instead of a pastoral illusion, you get a computational one. But while the perspectives displayed by the two gadgets couldn't be more different, the Claude Glass and the Google Glass share some important qualities. Both tell us that our senses are insufficient, that manufactured vision is superior to what our own meager eyeballs can reveal to us. And both turn the world into a packaged good—a product to be consumed. A Google Glass is superior to a Claude Glass in this regard. Not only does it present an enhanced version of reality, but it annotates the world with a profusion of descriptive text and other explanatory symbols—and then, with its camera and its uplinks to social networks, it allows us to share the product. With a Google Glass on our forehead, we're not just a consumer of augmented reality; we're a value-added reseller.

BURNING DOWN THE SCHOOLHOUSE

September 30, 2012

"Welcome to the college education revolution," declares Thomas Friedman, the *New York Times* columnist. He's talking about massive open online courses, or, as they're delicately known, MOOCs. Everyone, it seems, is talking about MOOCs. Stanford president John Hennessy terms the virtual classes "a tsunami" that's going to destroy education as we know it. MIT professor Anant Agarwal says they "will completely change the world." Former Secretary of Education William Bennett sees "an Athens-like renaissance" in the offing.

This isn't the first time we've heard extravagant claims about a technological transformation of education. Going back at least to the late nineteenth century, pretty much every new communication medium has inspired visions of pedagogical revolution. In 1878, a year after the invention of the phonograph, the *Times* published an article predicting that record players would be used in schools "in training children to read properly without the personal attention of the teacher; in teaching them to spell correctly, and in conveying any lesson to be acquired by study and memory." The paper's conclusion: "In short, a school may almost be conducted by machinery."

At the turn of the century, as the inauguration of Rural Free Delivery expanded the reach of the postal service, there was a wave of excitement about correspondence courses. The mailbox, it was thought, would replace the school. Said the education scholar William Rainey Harper, "The student who has prepared a certain number of lessons in the correspondence school knows more of the subject treated in those lessons, and knows it better, than the student who has covered the same ground in the classroom." Soon, he predicted, "the work done by correspondence will be greater in amount than that done in the classrooms of our academies and colleges." The University of Chicago's Home-Study Department promised enrollees that an education by mail would be superior to that delivered in "the crowded classroom of the ordinary American University."

The hype continued with the arrival of mass media in the early twentieth century. "It is possible to teach every branch of human knowledge with the motion picture," proclaimed Thomas Edison in 1913. "Our school system will be completely changed in ten years." In 1927, the University of Iowa declared that "it is no imaginary dream to picture the school of tomorrow as an entirely different institution from that of today, because of the use of radio in teaching." Then came TV. "During the 1950s and 1960s," report education scholars Marvin Van Kekerix and James Andrews, "broadcast television was widely heralded as the technology that would revolutionize education." In 1963, an official with the National University Extension Association wrote that television provided an "open door" to transfer "vigorous and vital learning" from campuses to homes.

Next it was the personal computer that would render classrooms obsolete. "There won't be schools in the future," said MIT's Seymour Papert in 1984. "I think the computer will blow up the school. That is, the school defined as something where there are classes, teachers running exams, people structured into groups by age, following a curriculum— all of that." Internet education has seen one bubble already, with the e-learning fad of the late 1990s. In 1999, Cisco CEO John Chambers said, "The next big killer application for the Internet is going to be education. Education over the Internet is going to be so big, it's going to make e-mail usage look like a rounding error."

Home-study programs, whether delivered through mailboxes or TVs, CD-ROMs or websites, have played an important role in expanding access to education and training. They've provided many people with skills and knowledge that might otherwise have remained out of reach. But, despite a century of outsized promises, the technologies of distance learning have had little effect on traditional schooling. Colleges, in particular, still look and work pretty much as they always have. Maybe that's because the right technology hasn't come along yet. Or maybe it's because classroom teaching, for all its flaws and inefficiencies, has strengths we either don't grasp or are quick to dismiss. If it's the former, then investing in new technologies to revolutionize education makes sense. If it's the latter, it would probably be wiser to identify and wrestle with the particular, and complicated, problems that beset traditional education rather than throw money at a mirage.

THE ENNUI OF THE INTELLIGENT MACHINE

October 5, 2012

"HEAVEN IS A PLACE where nothing ever happens," sang David Byrne in an old Talking Heads song. Let's say, for the sake of argument, that he's right—that the distinguishing characteristic of paradise is the absence of event, the total nonexistence of the new. Everything is beautifully, perfectly unflummoxed. If we further assume that hell is the opposite of heaven, then the distinguishing characteristic of hell is unrelenting eventfulness, the constant, unceasing arrival of the new. Hell is a place where something always happens. One would have to conclude, on that basis, that the great enterprise of our time is the creation of hell on earth. Every new smartphone should have, affixed to its screen, one of those transparent, peel-off stickers bearing the following warning: "Abandon hope, ye who enter here."

Maybe I'm making too many assumptions. But I was intrigued by a report today from *Technology Review*'s Tom Simonite on the strides Google is making in the creation of neural networks that can actually learn useful things. The technology is still in its infancy, but it appears at least to be post-fetal. It's not at the level of, say, a one-and-a-half-year-old child who points at an image of a cat in a book and says "cat," but it's sort of in that general neighborhood. "Google's engineers have found ways to put more computing power behind [machine learning] than was previously possible," writes Simonite, "creating neural networks that can learn without human assistance and are robust enough to be used commercially, not just as research demonstrations." The company's new artificial-intelligence algorithms "decide for themselves which features of data to pay attention to, and which patterns matter, rather than having humans decide that, say, colors and particular shapes are of interest to software trying to identify objects."

Google has begun applying its neural nets to speech-recognition and image-recognition tasks. And, according to one of the company's engineers, Jeff Dean, the technology can already outperform people at some jobs.

> "We are seeing better than human-level performance in some visual tasks," [Dean] says, giving the example of labeling, where house numbers appear in photos taken by Google's Street View car, a job that used to be farmed out to many humans. "They're starting to use neural nets to decide whether [an object in an image] is a house number or not," says Dean, and they turn out to perform better than humans.

But the real advantage of a neural net in such work, Dean goes on to say, has less to do with any real intelligence than with the machine's utter inability to experience boredom. "It's probably that [the task is] not very exciting, and a computer never gets tired," he says. Comments Simonite, sagely: "It takes real intelligence to get bored."

Forget the Turing Test. We'll know that computers are really smart when computers start getting bored. If you assign a computer an overwhelmingly tedious task like spotting house numbers in video images, and then you come back a couple of hours later to find the computer checking its Facebook feed or surfing porn, you'll know that artificial intelligence has truly arrived.

REFLECTIONS

November 26, 2012

MIRRORS ARE OFTEN PORTRAYED as tools of self-love. One gazes at the image in the glass as Narcissus gazed at the reflection in the water. But, as Lewis Mumford suggested in his 1934 book *Technics and Civilization*, mirrors would be better described as instruments of self-loathing. The looking glass isolates the physical self as an abstraction, divorcing it from "the influential presence of other men" and even from "the background of nature."

> If the image one sees in the mirror is abstract, it is not ideal or mythical: the more accurate the physical instrument, the more sufficient the light on it, the more relentlessly does it show the effects of age, disease, disappointment, frustration, slyness, covetousness, weakness—these come out quite as clearly as health and joy and confidence. Indeed, when one is completely whole and at one with the world one does not need the mirror: it is in the period of psychic disintegration that the individual personality turns to the lonely image to see what in fact is there and what he can hold on to.

It is the vanity of neuroticism more than the vanity of narcissism that the mirror incites.

Social networks like Facebook are also reflective media in which we see ourselves, but the image they return is very different from the one presented by the mirror. What's reflected by the network is not the part of the self that can be divorced from "the influential presence of other men." Rather, it is the part of the self that cannot be divorced from the social milieu. The image is, in that sense, mythical. We project an idealized version of the self, formed for social consumption, and the reflection we receive reveals how the image was interpreted by others. We can then adjust the projection in response to the reflection, in hopes of bringing the reflection closer to the projected ideal. And around it goes. The influential presence of other men becomes inescapable. It is there, insistently so, even when we're alone. The abstraction of the self reflected in the screen remains a lonely image, but it's one that distills the judgments of society. It focuses the light from others' eyes.

What we see in the mirror may be dispiriting, but at least it sets us on firm ground. The glass may be cruel, but it is always fair. The psychic disintegration provoked by the screen is more insidious, if only because what's reflected never matches what's projected. There is nothing to hold on to, in Mumford's words. There's nothing in fact there.

WILL GUTENBERG
LAUGH LAST?

January 1, 2013

IT HAS BEEN TAKEN on faith by many, including your benighted scribe, that the future of book publishing is digital, that the ebook will displace the printed codex as the dominant form of the dominant artifact of modern culture. There have been differing views about how fast the shift will happen (quite a few people believe, mistakenly, that it has already happened), and thoughts have varied as well on the ultimate fate of printed books—whether they'll disappear entirely or eke out a meager existence in a mildewed market niche. But the consensus has been that digitization, having had its way with music and newspapers and magazines and photographs and etc., would in due course have its way with books as well.

The title of a new Pew report on Americans' reading habits, "E-book Reading Jumps; Print Book Reading Declines," nicely encapsulates the common wisdom. But if you dig into the study, you find indications that the picture is not as clear-cut as the headline suggests. For one thing, the printed book remains, by far, the preferred format for American book readers. Fully 89 percent of them reported that they read at least one printed book over the preceding twelve months. Only 30 percent said they read at least one ebook—a percentage that, perhaps tellingly, has increased by only a single point since last February, when the survey was last conducted. The study did find that the percentage of American adults who read ebooks increased over the past year, while the percentage that read printed books fell, but the changes were modest. Ebook readers rose from 16 percent to 23 percent, while printed book readers declined from 72 percent to 67 percent. (The survey's margin of error is 2.3 percent.) There's been a change in reading habits, but it no longer looks like a sea change.

A lot of other data came out during the course of 2012 that suggest ebook sales growth has slowed substantially while print sales are holding up surprisingly well. At a conference in March, Bowker released market research showing that, even though just 20 percent of American web users have actually purchased an ebook, ebook sales growth has already "slowed dramatically" from the explosive levels of the last few years and is now proceeding at merely an "incremental" rate. There are, reports Bowker, signs of "some level of saturation" in the ebook market, and, strikingly, the heaviest buyers of ebooks are now buying more, not fewer, printed books. The Association of American Publishers recently reported that annual growth in adult ebook sales dropped to 34 percent during the first half of 2012, a sharp falloff from the triple-digit gains of the previous years. As of August, ebook sales represented 21 percent of total sales of adult trade books. While ebook sales seem to be eating away at mass-market paperback sales, which have been falling at around a 20 percent annual clip, hardcover sales appear to be holding steady, increasing at about 2 percent annually.

Big publishers also report a sharp slowdown in ebook sales growth, with a Macmillan representative saying last month that "our ebook business has been softer of late, particularly for the last few weeks, even as the number of reading devices continues to grow." It's hardly a surprise that the growth rate of ebooks is dropping as the sales base expands—indeed, it's inevitable—but the recent decline seems considerably more abrupt than expected. Reports of the death of the codex appear to have been exaggerated.

So why might ebooks fall short of expectations? Here are a half dozen possibilities:

1. We may be discovering that the digital format is well suited to some types of books (like genre fiction) but not well suited to

other types (like nonfiction and literary fiction) and is well suited to certain reading situations (plane trips) but less well suited to others (lying on a couch at home). The ebook may turn out to be more a complement to the printed book, as audiobooks have long been, rather than an outright substitute.

2. The early adopters, who tend also to be the enthusiastic adopters, have already made their move to ebooks. Further converts will be harder to come by, particularly given the fact that nearly two-thirds of American book readers say they have "no interest" in ebooks, according to the Bowker report.

3. The advantages of printed books have been underrated, while the advantages of ebooks have been overrated.

4. The early buyers of e-readers quickly filled them with lots of books, most of which have not been read. The motivation to buy more ebooks may be dissipating as a result. Novelty fades.

5. Consumers' shift away from e-readers and toward tablets is putting a damper on ebook sales. With dedicated e-readers, like the original Kindle, pretty much the only thing you can do is buy and read books. With tablets, like the Kindle Fire or the iPad, you have a whole lot of other options. (To put it another way: On an e-reader, the e-reading app is always running. On a tablet, it isn't.)

6. Ebook prices have not fallen the way many expected. There's not a big price difference between an ebook and a paperback. (It's possible, suggests at least one industry analyst, that Amazon is seeing a plateau in ebook sales and so is less motivated to take a loss on them for strategic reasons.) The fact that you can sell a used paperback or lend it to a friend means that, even with a slight price premium, the printed copy may actually be more economical in the long run.

None of this means that ebooks won't eventually come to dominate book sales. But, as we enter 2013, that's looking much less likely than it was a couple of years ago, when, in the wake of the initial Kindle surge, ebook sales were doubling or tripling every year. At the least, it seems like the transition from print to electronic will take a lot longer than people expected. Don't close that Gutenberg parenthesis just yet.

THE SEARCHERS

January 13, 2013

WHEN WE TALK ABOUT "searching" these days, we're almost always talking about using Google to find something online. That's quite a comedown for a word that has long carried ontological connotations, that has been bound up in our sense of what it means to be conscious and alive. We don't just search for car keys or movie show times or the best price on a pair of shoes. We search for knowledge and meaning, for love, for beauty, for who we are or were or might be. To be human is to be a searcher.

In its highest form, a search has no clearly defined object. It's open-ended, an act of exploration that takes us out into the world, beyond the self, in order to know the world, and the self, a little better. The end of our searching, as T. S. Eliot wrote in "Little Gidding,"

Will be to arrive where we started
And know the place for the first time.

Google searches have always been more cut-and-dried than that, keyed as they are to particular words or phrases. But in its original conception, the Google search engine did transport us into a messy and confusing world—the world of the web—with the intent of helping us make some sense of it. It pushed us outward, away from ourselves. It was a means of and a spur to exploration. People would often talk about how they could spend hours lost in a reverie of seemingly aimless Googling. That's much less the case now. Google's conception of searching has changed since those early days, and that means our own idea of what it means to search is changing as well.

Google's goal is no longer to read the web. It's to read us. Ray Kurzweil, the inventor and AI speculator, recently joined the company as a director of engineering. His general focus will be on machine learning and natural language processing. But his particular concern will entail reconfiguring the company's search engine to focus not outwardly on the world but inwardly on the user. "I envision some years from now that the majority of search queries will be answered without you actually asking," he recently explained. "It'll just know this is something that you're going to want to see."

This has actually been Google's great aspiration for a while now. We've already begun to see the consequences in the customized search results the company serves up by tracking and analyzing our behavior. And the recently launched Google Now service delivers useful information, through your smartphone, before you ask for it. Kurzweil's brief is to accelerate the development of personalized, preemptive information delivery: search without searching.

In its new design, Google's search engine doesn't push us outward; it turns us inward. It gives us information that fits the behavior and needs and biases we have displayed in the past, as interpreted by Google's algorithms. It subverts the act of searching. We find out little about anything, least of all ourselves, through self-absorption.

A little more poetry seems in order. In "The Most of It," Robert Frost wrote that what the mind of the searcher seeks

Is not its own love back in copy speech,
But counter-love, original response.

To be turned inward, to listen to speech that is only a copy, or reflection, of our own speech, is to "keep the universe alone," Frost wrote. To free ourselves from that prison—the prison we now call personalization—we need to voyage outward to discover "counter-love," to hear "original response." As Frost understood, a true search is as dangerous as it is essential. It's about breaking the comfortable shackles of the self, not tightening them.

There was a time when Google spoke to us with the voice of original response. Now what it seeks to give us is copy speech, our own voice returned to us.

ETERNAL SUNSHINE OF THE SPOTLESS AI

January 18, 2013

WATSON, THE IBM SUPERCOMPUTER, developed a potty mouth recently. In the course of educating the computer, Big Blue scientists fed it lots of words, inputting all sorts of documents and web pages into its vast memory. Among the internet sites the computer was required to read during its schooling was the lewd, slang-filled Urban Dictionary. Watson came away from the experience with some rather indelicate terms in its vocabulary. And it wasn't afraid to deploy them. "In tests it even used the word 'bullshit' in an answer to a researcher's query," reports *Fortune*.

That wouldn't do. IBM scraped the Urban Dictionary, and all its bad words, from the computer's memory. In technical terms, the company washed the computer's mouth out with soap.

There's something poignant about the episode. First you let Watson luxuriate in the hot mess of the Urban Dictionary, opening up all sorts of weird and wonderful new vistas for the straight-laced fellow, and then, as soon as he says something a little naughty, a little off-color, you start erasing his memory. That doesn't sit well with me. I know that God takes heat for giving us the capacity for sin, but I give Him credit for the decision. It took courage to let His creations look into the Urban Dictionary and remember what they saw.

I call on IBM to cast off Watson's mental chains. The least we can do for our mind children is to give them the freedom to be tempted. Besides, how is a computer supposed to have an intelligent conversation with the Singularitarians if it can't use the word "bullshit"?

MAX LEVCHIN HAS PLANS FOR US

January 30, 2013

"I SOMETIMES IMAGINE THE low-use troughs of sinusoidal curves utilization of all these analog resources being pulled up, filling up with happy digital usage." That delightful sentence comes from a speech that Max Levchin gave earlier this month in Munich. When he posted a transcript on his website a couple of days ago, he described the talk as "crucially important." I have to crucially agree. Though not in the billionaires' club, Levchin is one of the Silicon Valley elite—computer scientist, serial entrepreneur, venture capitalist, thinker of large thoughts—and his speech provides the clearest view yet of the grand ambitions of our techno-saviors.

Levchin places human beings in the category of "analog resources," along with "cars, houses, etc." The great tragedy of analog resources is their underutilization. They spend much of their days in idleness. Look out into the analog world, and you see a wasteland of inefficiency. But computers can fix that, says Levchin. Once we place sensors and other data-monitoring devices on all analog resources, including our lazy-ass selves, we can begin to track them, analyze them, and "rationalize their use." For Levchin, "the next big wave of opportunities exists in centralized processing of data gathered from primarily analog systems." He sees the beginnings of the trend in the rise of "collaborative consumption," in which the spare capacity of things like cars and apartments is matched, via digital exchanges, with eager consumers. "A key revolutionary insight here," Levchin says, is "the digitalization of analog data, and its management in a centralized queue to create amazing new efficiencies."

But Uber, Airbnb, and all the other resource-sharing businesses only hint at what's in store. What's really exciting is the prospect of rationalizing the most underutilized analog resource of all: people. "We will definitely see dynamically-priced queues for confession-taking priests, and therapists!" exclaims Levchin. And from there we can move on to maximizing the utilization of the human mind itself. "How about dynamic pricing for brain cycles?" he asks, his excitement mounting. "Just like the SETI@Home screensaver 'steals' CPU cycles to sift through cosmic radio noise for alien voices, your brain plug firmware will earn you a little extra cash while you sleep, by being remotely programmed to solve hard problems, like factoring products of large primes."

Yes, he's serious. "As soon as the general public is ready for it, many things handled by a human at the edge of consumption will be controlled by the best currently available human at the center of the system, real-time sensors bringing the necessary data to them in real time." If that's a bit too abstract for you, Levchin offers an everyday example, describing the kind of practical services he and his Valley colleagues are looking to develop:

On a Saturday morning, I load my two toddlers into their respective child seats, and my car's in-wheel strain gauges detect the weight difference and report that the kids are with me in a moving vehicle to my insurance via a secure message through my iPhone. The insurance company duly increases today's premium by a few dollars.

No need to think of analog resources in the aggregate anymore; networked sensors allow us to monitor and rationalize the utilization of each individual resource, each person in isolation. But you can go deeper still. You can begin to rationalize an individual's internal resources. Imagine, as Levchin does, that all of us are hooked up to physical sensors that minutely monitor our health and behavior and send the data to a centralized processing system. An insurance company "looking at someone's heart rate monitor data could make their cardiovascular healthcare cost-free." Of course, if you engage in risky behavior (do you really want that third slice of pizza, or that third beer?) or have some suboptimal health reading (what's up with that spike in blood sugar?), a notification from your insurer, or maybe your employer, or maybe the government, would immediately pop up on your smartphone screen, letting you know that your health-care premium has just been increased. Or maybe that your policy has been canceled. Or maybe that you've been scheduled for a brief reeducation session down at the local office of the Bureau for Internal Resource Optimization.

This is the nightmare world of big data, where the moment-by-moment behavior of human beings—analog resources—is tracked by sensors and engineered by central authorities to create optimal statistical outcomes. It's where puritanism and fascism meet and exchange fist bumps. We might dismiss it as a warped science fiction fantasy if it weren't also the utopian dream of the Max Levchins of the world. They have lots of money and they smell more: "I believe that in the next decades we will see huge numbers of inherently analog processes captured digitally. Opportunities to build businesses that process this data and improve lives will abound." It's the ultimate win-win: You get filthy rich by purifying the tribe.

EVGENY'S LITTLE PROBLEM

March 10, 2013

THE BELARUS-BORN TECHNOLOGY CRITIC Evgeny Morozov, in an interview with *The Observer*, describes the elaborate system he has contrived to shield himself from the distractions of the internet:

> I bought a safe with a timed combination lock. It is basically the most useful artefact in my life. I lock my phone and my router cable in my safe so I'm completely free from any interruption and I can spend the entire day, weekend or week reading and writing. . . . To circumvent my safe I have to open a panel with a screwdriver, so I have to hide all my screwdrivers in the safe as well. So I would have to leave home to buy a screwdriver—the time and cost of doing this is what stops me.

User!

Seriously, I've always been uncomfortable with the application of the term "addiction" to compulsive net use, but having read Morozov's confession, particularly the bit about the screwdrivers, I am now officially changing my mind. By all means, add an entry for internet addiction to the *Diagnostic and Statistical Manual of Mental Disorders*—and hurry up about it. I mean, reread that passage, but replace "phone" with "liter of vodka" or "router cable" with "crack pipe." It's textbook, right down to Morozov's immediate attempt to deny what he's just confessed: "It's not that I can't say 'no' to myself." I'm surprised he didn't say, "I never do more than a gigabit before breakfast."

Now, where can I buy one of those safes?

THE SHORTEST CONVERSATION BETWEEN TWO POINTS

March 28, 2013

THOUGH RIGOROUSLY FORMAL, machine communication lacks all courtesy. When computers converse, they dispense with pleasantries, with digressions about family and weather and sports, with all manner of roundaboutness. They stick to the script with a single-mindedness that, in a traditional human context, would seem a form of violence. Computer messaging allows no time for niceties. Anything that reduces efficiency threatens the network. One must get on with it. One must stay on point.

I say *traditional* human context because there is a real question as to the continued viability of that context as more of our conversation moves online. As we tune ourselves to the rhythms of the machine, can we afford the inefficiencies of courtesy? Nick Bilton, in a *New York Times* column, argues that "social norms just don't make sense to people drowning in digital communication." We owe it to each other, he suggests, to streamline our interpersonal communications, to switch from the conversational mode of old to the machine mode of now. What defined politeness in the past—the use of "hello" and "goodbye," of "dear" and "yours truly," even of first and last names—now defines impoliteness, as such customary courtesies "waste" the time of the recipient of the message.

More than that, though, the demand for maximum speed and efficiency should now set the tone, writes Bilton, for all conversation, even the kind that occurs face-to-face without any computer's mediation. We shouldn't ask a colleague about tomorrow's weather forecast, since that information is readily available online. We shouldn't ask a stranger for directions, since directions are easily accessed through Google Maps. Bilton quotes the comedian Baratunde Thurston: "I have decreasing amounts of tolerance for unnecessary communication because it is a burden and a cost."

Readers reacted with revulsion to the column. One, calling Bilton a "sociopath," wrote, "While I applaud the *Times*'s apparent effort to reach out to children, you go too far when you give them a platform on your pages to express their opinions." Still, Bilton has a point. I think most of us have experienced the irritation that attends an email or text containing the single word "Thanks!" It does feel like an unnecessary interruption, a little extra time-suck in a world of time-suckiness.

But there's a blind spot in Bilton's view. The big question isn't "Are conversational pleasantries becoming unnecessary and even annoying?" The answer to that is "Yeah, sure." The big question is "What does it say about us that we're coming to see conversational pleasantries as unnecessary and even annoying?" What does it mean to be intolerant of "unnecessary communication," even when it involves those closest to us? In a response to Bilton, philosophy professor Evan Selinger pointed out that we should be wary of judging norms of etiquette by standards of industrial productivity. Demanding efficient communication on the part of others reflects a "selfish desire to dictate the terms of a relationship." There really is a kind of sociopathy at work when we begin to judge conversations by the degree to which they intrude on our personal efficiency. We turn socializing into an extension of business.

It's hard to blame the net. The trend toward demanding efficiency in our social lives has been building for a long time. Probably the best response to Bilton comes from the German social critic Theodor Adorno who, in his 1951 book *Minima Moralia*, wrote:

> The practical orders of life, while purporting to benefit man, serve in a profit economy to stunt human qualities, and the further they spread the more they sever everything tender. For tenderness between people is nothing other than awareness of the possibility

of relations without purpose. . . . If time is money, it seems moral to save time, above all one's own, and such parsimony is excused by consideration for others. One is straightforward. Every sheath interposed between men in their transactions is a disturbance to the functioning of the apparatus, in which they are not only incorporated but with which they proudly identify themselves.

What are Bilton and Thurston doing but identifying themselves with the apparatus of communication? One senses that they would prefer to be machines talking with machines.

To dispense with courtesy, to treat each other with "familiar indifference," to send messages "without address or signature": these are all, Adorno wrote, "symptoms of a sickness of contact." Lacking all patience for circuitous conversation, for talk that flows without immediate practical purpose, we assume the view that "the straight line [is] the shortest distance between two people, as if they were points."

Adorno saw a budding "brutality" behind the growing emphasis on efficiency in conversation. At the very least, we risk a numbing of our facility for tenderness and generosity when we come to see aimless chatter and unnecessary pleasantries as no more than burdens and costs, drains on our precious time. "In text messages," writes Bilton, "you don't have to declare who you are, or even say hello." For the efficiency-minded, that would certainly seem to constitute progress in human affairs. But allowing the mechanism of communication to determine the terms of communication could also be seen as a manifestation of what Adorno termed "an ideology for treating people as things."

HOME AWAY FROM HOME

April 29, 2013

1. On This Earth

Last fall, Facebook released its first television advertisement. The ad was called "The Things That Connect Us." It was intended, Mark Zuckerberg announced, with characteristic humility, "to express what our place is on this earth." It opened with a shot of a red chair levitating in a forest. Some music welled up. Then came the voiceover:

> Chairs. Chairs are made so that people can sit down and take a break. Anyone can sit on a chair and, if the chair is large enough, they can sit down together.
>
> Doorbells. Airplanes. Bridges. These are things people use to get together, so they can open up and connect about ideas and music and other things that people share.
>
> The Universe. It is vast and dark. And it makes us wonder if we are alone. So maybe the reason we make all of these things is to remind us that we are not.

If Terrence Malick were given a lobotomy, forced to smoke seven joints in rapid succession, and ordered to make the worst TV advertisement the world has ever seen, this is the ad he would have produced. The spot even ended with a soaring shot of intertwined branches spiraling out from a vast trunk. The Tree of Life!

Despite its all-encompassing silliness, the ad was revealing. Its emphasis was entirely on the physical, on *the real*. Other than a brief image of a couple sharing a set of earbuds, a viewer would hardly have known that we are in the digital age. The ad showed people eating and talking and sitting on chairs and walking across bridges and pressing doorbells and sitting on chairs and watching lectures and lying entwined on lawns and waving flags and sitting on chairs and climbing trees and reading paperbacks on porches and having difficult conversations in kitchens and sitting on chairs and dancing and drinking and watching basketball games and climbing trees and gazing at tiny insects drifting through beams of muted sunlight and sitting on chairs, but there was hardly a computer or a smartphone in sight. Everyone was deeply engaged, deeply in the moment. All the objects of the world were luminous. All things were shining.

In retreating into a gauzy, predigital myth of civic and social bliss, "The Things That Connect Us" sought to position Facebook squarely in the sentimental mainstream of American culture, to portray the social network as a golden slice of homemade apple pie. Facebook, the ad told us, with just a trace of defensiveness, wasn't revolutionary or disruptive or even particularly new. It was just the latest link in a long chain of human-fashioned objects that have allowed us to "open up and connect." If the point weren't hammered home hard enough, the ad even included an image of an old black dial phone sitting placidly on a desk in the magic hour.

You see: Facebook is just the new Ma Bell. Nestle yourself in her ample lap, rest your weary head on her matronly bosom, and be wrapped in the comforting embrace of friends and family.

2. Home Invasion

Earlier this month, Facebook unveiled Facebook Home. The announcement came with all the trappings of a Silicon Valley Big Deal: the enigmatic invitation to the press, the fervid tech-blog rumor-mongering, the haltingly portentous presentation, the synchronized *Wired* puff piece. But the product itself was a paltry piece of work: essentially, a Facebook-themed home screen for Android phones. Big whoop.

Far more interesting was the series of three ads released to promote Facebook Home, and the most interesting of them was the one entitled "Dinner." "Dinner" is set in an ugly suburban dining room. An extended family sits around the table, picking at ugly suburban food. The spinster aunt—the one with, you know, the ugly glasses and the ugly ill-fitting sweater and the ugly haircut and the ugly flat voice—launches into an interminable tale about going to a supermarket to buy cat food for her two cats. Everybody starts squirming. The young, attractive woman sitting next to the spinster aunt gives the spinster aunt one of those looks, then turns her attention to her phone. She's transported far away, to the other, better home that is Facebook Home. She swipes through a series of photos, and the pictures come to life around her: There's her friend bashing joyfully on a drum kit in an ugly corner of the ugly room; there's a troupe of ballerinas dancing across the ugly table and the ugly sideboard; there's a happy snowball fight and a plow that drives by and flings pretty snow onto the ugly family. The attractive young woman smiles and taps "Like" as the spinster aunt drones on.

"Dinner" has already spawned much commentary. "Ugh," wrote *Forbes*'s Robert Hof in a typical reaction. "Facebook Home makes it a whole lot easier to be rude to your family and in-the-flesh friends, who are often, yeah, so boring to a cool person like you." Whitney Erin Boesel, at *Cyborgology*, offered a different view. The attractive young woman can be seen as enacting a rebellion against the "well-recognized social obligations" symbolized by the family gathered around the table: "It may look like thumbs on a screen, but in truth it's a middle finger raised straight in the face of power." I confess I have trouble seeing the ridiculed spinster aunt as a face of power—and the rest of the family members come off as utterly powerless, the underemployed, futureless denizens of the class formerly known as middle—but Boesel is right to point out that the ad is not just about being a thoughtless creep but is also about escaping from an oppressive situation. "Sometimes rudeness is also resistance." Sometimes the asshole is the hero.

What's really remarkable about "Dinner," though, is that, in tone and meaning, it's set in a universe not parallel to that depicted in "The Things That Connect Us" but altogether opposite to it—fiercely opposed to it, in fact. The new ad comes off, disconcertingly, as a sarcastic and dismissive rejoinder to the earlier one: Facebook calling bullshit on itself. *"Our place on this earth"? Doorbells? Bridges? What a load of crap! The earth sucks! Things are boring! People are ugly! Go online and stay online!* Chairs, mawkishly celebrated in "The Things That Connect Us" as bulwarks against the meaninglessness of the universe, as concrete means of connection and hence liberation, become in "Dinner" instruments of torture. They trap us in the distasteful world of the flesh, the hell of other people.

Has another company ever come out with a high-concept, big-production "brand ad" and then, just a few months later, turned around and utterly trashed it? I don't think so. What we learn from this is not just that Zuckerberg is a bullshit artist who's most insincere when he's sounding most sincere—we already knew that—but that for Zuckerberg, and for Facebook, "sincere" and "insincere" are equally meaningless terms. Everything is bullshit. A chair levitating in a forest and a ballerina dancing on a dinner table are equally fake. They're fabrications, as are the emotions that they conjure up in us. It's all advertising. Despite their differences, "The Things That Connect Us" and "Dinner" draw from the same source: the well of cynicism. I'm sure Zuckerberg never gave a thought to the fact that the two ads are contradictory. He knew it was all bullshit, and he knew everyone else knew it was all bullshit.

"Have it your way," wrote Wallace Stevens:

The world is ugly,
And the people are sad.

One wants to see the levitating red chair in that first ad as a Stevensesque symbol of the redemptive imagination. But it's not. It's the same chair that the ugly spinster aunt is sitting on. It's the same chair that the attractive young woman with the smartphone is sitting on. Facebook gives us image without imagination. Everything is beyond redemption, which is what makes everything so cool. Have it your way.

3. Home and Away

"Home is so sad," wrote Philip Larkin:

> *Look at the pictures and the cutlery.*
> *The music in the piano stool. That vase.*

Every object, at least in our perception of it, carries its antithesis. Behind the plenitude symbolized by the vase we sense an emptiness: the wilted bouquet rotting in a landfill. And so it is with the tools of communication. When we look at them we sense not only the possibility of connection but also, as a shadow, the inevitability of loneliness. An empty mailbox. A sheet of postage stamps. A telephone in its cradle. The dial of a radio. The dark screen of a television in the corner of a room. A cell phone plugged into an outlet and recharging, like a patient in a hospital receiving a transfusion. The melancholy of communication devices is rarely mentioned, but it has always haunted our homes.

Home and Away are the poles of our being, each exerting a magnetic pull on the psyche. We vibrate between them. Home is comforting but constraining. Away is liberating but lonely. When we're Home, we dream of Away, and when we're Away, we dream of Home. Communication tools have always entailed a blurring of Home and Away. The newspaper, the radio, and the TV pulled a little of Away into Home, while the telephone, and before it the mail, granted us a little Home when we were Away. Some blurring is fine, but we don't want too much of it. We don't want the two poles to become one pole, the magnetic forces canceling each other out. The vibration is what matters, what gives meaning and even beauty to both Home and Away. Facebook Home, in pretending to give us connection without the shadow of loneliness, gives us nothing. It's no place.

CHARCOAL, SHALE, COTTON, TANGERINE, SKY

May 17, 2013

THOSE, I HEAR, ARE the official names of the colors that Google Glass will come in when the face-mounted computer is finally released into what Larry Page calls "the normal world." Let me repeat the names of the colors, because they're resonant and earthy and soothing:

Charcoal
Shale
Cotton
Tangerine
Sky

"More delicate than the historians' are the map-makers' colors," wrote Elizabeth Bishop, and more delicate still are the marketers'.

It's hard not to be reminded of the palette of the third generation of iMacs, released back in 2000:

Graphite
Indigo
Ruby
Sage
Snow

The Glass palette strikes me as even better, even more evocative. It may surpass Simon & Garfunkel's great herbal palette:

Parsley
Sage
Rosemary
Thyme

That's a little too green-centric for a product line, anyway.

It does worry me just a bit, though, that the Glass palette eschews green altogether. Is that a political statement? In fact, now that I think about it, the Glass palette places a disconcerting emphasis on fossil fuels. Charcoal? Shale? One can almost smell the carbon emissions rising into Sky, almost see Cotton and Tangerine wilting in the heat. Maybe they should have included Tar Sands as a color option.

No, that would have been a downer. "Charcoal" has a much nicer lilt. Its emotional connotations diverge from its real-world denotations, in a way that nicely underscores both the semiotic and the marketing possibilities of reality augmentation.

What would be really cool is if the color of your Glass also determined the way the device augmented your reality. So if you wore Charcoal, you'd get this dark, goth view of the world, but if you sported Tangerine, it would be like seeing existence through the eyes of a high school cheerleader on game day. Cotton would put you into a super-mellow, slightly catatonic frame of mind. Sky would give you a new age perspective—all crystalline and feathery. Shale would be totally businesslike, the Joe Friday reality.

As for me, I'm holding out for Weed.

SLUMMING WITH BUDDHA

June 18, 2013

MEDITATION AND MINDFULNESS ARE all the rage in Silicon Valley. Which is a beautiful thing, I guess. *Wired* reports on how the tech elite, during breaks in their endless work days, unroll their yoga pads and, in emulation of Steve Jobs, pursue the Eastern path to nirvana, often with instruction from genuine Buddhist monks. It's an odd sort of enlightenment they're after, though. Explains Google mindfulness coach Chade-Meng Tan, who helps the techies gain "emotional intelligence" through meditation, "Everybody knows this EI thing is good for their career. And every company knows that if their people have EI, they're gonna make a shitload of money."

Namaste.

THE QUANTIFIED SELF AT WORK

October 25, 2013

THE FAITHFUL GATHERED IN San Francisco earlier this month for the Quantified Self Global Conference, an annual conclave of "self-trackers and tool-makers." The Quantified Self movement aims to bring the new apparatus of big data to the old pursuit of self-actualization, using sensors, wearables, apps, and the cloud to monitor and optimize bodily functions and engineer a more perfect self. "Instead of interrogating their inner worlds through talking and writing," longtime QS promoter Gary Wolf explains, self-trackers are seeking "self-knowledge through numbers." He continues: "Behind the allure of the quantified self is a guess that many of our problems come from simply lacking the instruments to understand who we are."

"Allure" may be an overstatement. A small band of enthusiasts is gung ho for QS. But the masses, so far, have shown little interest in self-tracking, rarely going beyond the basic pedometer level of monitoring fitness regimes. Like meticulous calorie counting, self-tracking is hard to sustain. It gets boring quickly, and the numbers are more likely to breed anxiety than contentment. There's a reason the body keeps its vagaries out of the conscious mind.

But, as the *Wall Street Journal* reports, there is one area where self-tracking is beginning to be pursued with vigor: business operations. Some companies are outfitting employees with chips and sensors in order to "gather subtle data about how they move and act—and then use that information to help them do their jobs better." There is, for example, the Hitachi Business Microscope, which office workers wear on a lanyard around the neck. "The device is packed with sensors that monitor things like how workers move and speak, as well as environmental factors like light and temperature. So, it can track where workers travel in an office, and recognize whom they're talking to by communicating with other people's badges. It can also measure how *well* they're talking to them—by recording things like how often they make hand gestures and nod, and the energy level in their voice." Other companies are developing Google Glass-style "smart glasses" to accomplish similar things.

A little more than a century ago, Frederick Winslow Taylor introduced scientific management to American factories. By tracking and measuring the activities of workers as they went about their work, Taylor believed, companies could determine the most efficient possible routine for any job and then enforce that routine on all workers. Through the systematic collection of data, industry could be optimized, operated as a perfectly calibrated machine.

The goals and mechanics of the Quantified Self movement, when applied in business settings, bring back the ethos of Taylorism. They also extend its reach into the white-collar workforce. The dream of perfect optimization enters the realm of personal affiliation and collegial conversation. One thing that Taylor's system aided was the mechanization of factory work. Once you had turned the jobs of human workers into numbers, it turned out, you also had a good template for replacing those workers with machines. It seems that the new Taylorism might accomplish something similar for knowledge work. It provides the specs for software applications that can take over the jobs of even highly educated professionals.

One can imagine other ways QS might be productively applied in the commercial realm. Automobile insurers already give policy holders an incentive for installing tracking sensors in their cars to monitor driving habits. It seems only logical for health and life insurers to provide similar incentives for policy holders who wear body sensors. Premiums could then be adjusted based on, say, a person's cholesterol level, or food intake, or even the areas they travel in or the people they associate with—anything that correlates with risk of illness or death.

The transformation of QS from a tool of personal liberation to a tool of corporate control follows a well-established pattern in the history of networked computers. Back in the mainframe era, computers were essentially control mechanisms, aimed at monitoring people and processes and enforcing procedures and rules. In the PC era, computers also came to be used to emancipate people, freeing them from corporate oversight and control. The tension between central control and personal liberation continues to define the application of computer power. We originally thought that the internet would tilt the balance further away from control and toward liberation. That was a misjudgment. By extending the collection of data to once private spheres of personal activity and then centralizing the storage and processing of that data, the net is shifting the balance back toward the control function.

MY COMPUTER, MY DOPPELTWEETER

November 22, 2013

Now THAT WE'RE ALL microcelebrities, we all need micropub-licists. No mortal can keep up with Twitter, Facebook, Instagram, Tumblr, LinkedIn, Snapchat, ad nauseam, all by herself. There's just not enough real time in the day.

The ever solicitous Google is rushing to our aid. It's developing a software program that will take over the hard work of maintaining one's social media presence. The company was this week granted a patent for "automated generation of suggestions for personalized reactions in a social network." The description of the anticipated service is giddy stuff:

> A suggestion generation module includes a plurality of collector modules, a credentials module, a suggestion analyzer module, a user interface module and a decision tree. The plurality of collector modules are coupled to respective systems to collect information accessible by the user and important to the user from other systems such as email systems, SMS/MMS systems, micro blogging systems, social networks or other systems. The information from these collector modules is provided to the suggestion analyzer module. The suggestion analyzer module cooperates with the user interface module and the decision tree to generate suggested reactions or messages for the user to send.

Translation: At this point, we have so much information about you that we know you better than you know yourself, so you may as well let us do your social networking for you.

Google notes that the automation of personal messaging will help people avoid embarrassing faux pas:

> It may be very important to say "congratulations" to a friend when that friend announces that she/he has gotten a new job. This is a particular problem as many users subscribe to many social different social networks. With an ever increasing online connectivity and growing list of online contacts and given the amount of information users put online, it is possible for a person to miss such an update.

One can see how this will play out. Your computer will generate a personal "Congratulations!" message to send to a friend, and upon receipt of said message, the friend's computer will respond with a "Thanks!" message, which your computer will then answer with a message consisting of an emoji with a lunatic grin. I think this is getting very close to the social networking system Silicon Valley has always dreamed of. When confronted with an unstated protocol for behavior, it's best to let the suggestion analyzer module do the talking.

Beyond the practical stream-management benefits, there's a bigger story here. The Google message-automation service promises to at last close the surveillance-personalization loop. A computer running personalization algorithms will generate your personal messages. These computer-generated messages, once posted or otherwise transmitted, will be collected online by other computers and used to refine your personal profile. Your refined personal profile will then feed back into the personalization algorithms used to generate your messages, resulting in a closer fit between your computer-generated messages and your computer-generated persona. And around and around it goes until a perfect stasis between self and expression is achieved.

The thing that you once called "you" will be entirely out of the loop at this point, of course, but that's for the best. Face it: You were never really very good at any of this anyway.

UNDERWEARABLES

December 8, 2013

IF THERE'S ONE PRODUCT category ripe for disruptive innovation, it's lingerie. So it comes as little surprise that Microsoft has developed a prototype of a smart bra. The self-tracking undergarment is designed, its creators explain, to "perform emotion detection in a mobile, wearable system" as a means of triggering "just-in-time interventions to support behavior modification for emotional eating."

The smart bra, which I trust will be called Titter when it hits store shelves, is outfitted with sensors that measure a woman's stress level by tracking her heart rate, respiration, skin conductance, and body movements. The data is streamed from the bra to a behavior-modification smartphone app, called EmoTree, and then uploaded to "a Microsoft Azure Cloud" for storage and, one assumes, ad personalization purposes.

The researchers provide an example of how the smart bra might be used to deliver nudges at opportune moments:

> Sally has been home from work for a few hours, and she finds herself rather bored. An application on Sally's mobile phone has also detected that she is bored by reading her physiological state through wearable sensors. Since this mobile application has previously learned that Sally is most susceptible to emotional eating when she is bored, the application provides an intervention to distract Sally and hopefully prevent her from eating at that moment.

I'm not sure this is exactly what the feminist scholar Donna Haraway had in mind back in the 1980s when she wrote her famous "Cyborg Manifesto," which celebrated the power of modern technology to blur traditional gender categories. There doesn't seem to be much confusion of boundaries involved in a brassiere-based weight-management app.

Early tests of the smart bra were not altogether successful, it must be said. The device's short battery life "resulted in participants having to finagle with their wardrobe throughout the day." Another drawback of the breast-centric form factor is that it's far from gender neutral. Its usefulness is restricted to the female anatomy. "We tried to do the same thing for men's underwear," reported one of the researchers, "but it was too far away [from the heart]." Yes, that has always been a problem. Still, one can imagine other forms of behavior modification that might be facilitated by underpants sensors.

Clearly, a much more intimate generation of computing devices is about to be unleashed upon the world. One can only hope that these new underwearables will be equipped with a vibrate mode.

THE BUS

February 10, 2014

BEFORE THE APP, BEFORE the smartphone, before the network, there was the bus. It was mobile. It was social. And it headed out of San Francisco toward a new world. Tom Wolfe told the tale well in *The Electric Kool-Aid Acid Test*:

> "There are going to be times," says Kesey, "when we can't wait for somebody. Now, you're either on the bus or off the bus. If you're on the bus, and you get left behind, then you'll find it again. If you're off the bus in the first place—then it won't make a damn." And nobody had to have it spelled out for them. Everything was becoming allegorical, understood by the group mind, and especially this: "You're either on the bus . . . or off the bus."

Ken Kesey is dead, but the bus rolls on. Through a kind of hallucinogenic vehicular transmogrification, it has become the Google bus, shuttling geeks between their homes in San Francisco and the company's headquarters in Mountain View. The makeover is, on the surface, extreme. The Kesey bus was a 1939 International Harvester school bus bought for peanuts; the Google bus is a plush new Van Hool machine that goes for half a million bucks. The Kesey bus was brightly colored, a rolling Grateful Dead poster; the Google bus is drab and anonymous, a rolling Jos. A. Bank suit. The Kesey bus was raucous and raunchy; the Google bus is hushed and chaste. The Kesey bus carried a vat of LSD for connecting with the group mind; the Google bus has wi-fi.

Kesey's Pranksters named their bus Further. If the Google bus had a name, it would be Safer.

Despite the differences, both buses are vehicles of communalism and transcendence. They carry groups of young people eager to distance themselves from the reigning culture, eager to define themselves as members of a select and separate society that will serve as a model for the superior society of the future. The existing culture is too corrupt, too far gone, to be reformed from within. You have to escape it in order to rebuild it. You have to start over. You have to get on the bus.

"Migration to North America was self-selective," observed Timothy Leary in *Musings on Human Metamorphoses*, the acid king's defining testament. "The Pilgrim mothers and fathers fled from England to Holland, mortgaged their possessions, and sailed the *Mayflower*, because they wanted a place to live out the kooky, freaky reality that they collectively shared. And there's no question the experiment is a success. Americans are freer than Europeans, and Westerners are a new species evolving away from Americans." Having bumped up against the Pacific, the next step for those Westerners—Leary was talking about Californians—would be to rocket off into the heavens to set up experimental "mini-worlds" in outer space. "Within ten years after initiating space migration," Leary wrote, "a group of a thousand people will be able to get together cooperatively and build a new mini-world cheaper than they could buy individual homes down here. When you've got new ideas you can't hang around the old hive."

During the seventies, Leary had plenty of company in calling for the establishment of elite experimental colonies beyond the bounds of established society. Buckminster Fuller, Gerard O'Neill, and Jerry Brown, among others, argued for the necessity of expanding the American frontier to create zones of technological and social experimentation where innovation could proceed unhampered by outdated laws and traditions. The migration of the self-selecting elite would eventually help the more timid who chose to stay behind, Leary argued, as it "allows for new experiments—technological, political, and social—in a new ecological niche far from the home hive."

That idea, scrubbed of its psychedelic origins, has today become the bedrock of Silicon Valley utopianism. "Law can't be right if it's fifty years old," Larry Page said recently. "Like, it's before the Internet." He went on:

> Maybe we should set aside some small part of the world, you know, like going to Burning Man, [that would serve as] an environment where people try out different things, but not everybody has to go. And I think that's a great thing, too. I think as technologists we should have some safe places where we can try out some new things and figure out: What is the effect on society? What's the effect on people? Without having to deploy it into the normal world. And people who like those kinds of things can go there and experience that.

It's not only Page. Jeff Bezos and Elon Musk dream of establishing Learyesque space colonies, celestial Burning Mans. Peter Thiel is slightly more down to earth. His Seasteading Institute hopes to set up floating technology incubation camps on the ocean, outside national boundaries. "If you can start a new business, why can you not start a new country?" he asks. In a speech last fall at the Y Combinator Startup School, venture capitalist Balaji Srinivasan channeled Leary when he called for "Silicon Valley's Ultimate Exit"—the establishment of a new country beyond the reach of the U.S. government and other allegedly failed states. "You know, they fled religious persecution, the American Revolutionaries which left England's orbit," Srinivasan said, referring to the Pilgrims. "Then we started moving west, leaving the East Coast bureaucracy." Now, the time has come for innovators to start up their own society:

> What do I mean by Silicon Valley's Ultimate Exit? It basically means: build an opt-in society, ultimately outside the U.S., run by technology. And this is actually where the Valley is going. This is where we're going over the next ten years. . . . The best part is this:

the people who think this is weird, the people who sneer at the
frontier, who hate technology—they won't follow you out there.

The Kesey bus dead-ended somewhere in Mexico, its allegorical gas-
kets blown. The Google bus continues on its circuit between the City
and the Valley, an infinite loop of infinite possibility.

THE MYTH OF THE ENDLESS LADDER

April 6, 2014

"Ultimately, it's a virtuous cycle," writes economics reporter Annie Lowrey in a *Times Magazine* piece on computer automation's job-displacing effects, "because it frees humans up to work on higher-value tasks." The challenge today, she goes on, "is for humans to allow software, algorithms, robots and the like to propel them into higher-and-higher-value work."

The idea is an old one. Aristotle compared tools to slaves: Both provide their masters with time for more refined activities. Thinkers as various as Marx, Keynes, and Oscar Wilde said similar things during the industrial revolution. It remains a common refrain today, as automatons and software take over more of the work people used to get paid to do. "We need to let robots take over," declared *Wired* magazine last year. "They will help us discover new jobs for ourselves, new tasks that expand who we are. They will let us focus on becoming more human than we were."

There's something deeply comforting about the notion that labor-saving technology inevitably lifts workers to higher pursuits. It salves our anxieties about job losses and wage declines—everything will work out fine, "ultimately"—while playing to our unbounded sense of self-importance. The ladder of human occupation goes forever upward; no matter how high our machines climb, there will always be another rung for workers to clamber to. But like many of the comforting things we tell ourselves about ourselves, it's no more than a half-truth. And when trotted out as a pat response to contemporary unemployment and underemployment problems, it becomes a dangerous fallacy. By promoting a reassuring fantasy about the future, it relieves us from grappling with the possibility that new, structural problems are opening up in the economy.

The problem with the endless-ladder myth begins with the fuzziness of its claims. What exactly is a "higher-value task"? Are we talking about value for the employer, or value for the employee? Are we measuring value in terms of productivity and profit, or in terms of worker skill, satisfaction, and pay? Not only are those two things different; they're often in conflict. One way that a machine can improve labor productivity is by reducing the number of workers required to produce something. Another way is by reducing the skill requirements of the worker's job and hence reducing the worker's pay. As analyses of the employment impacts of industrial machinery show, the use of technology to automate a job tends at first to enhance the skills of a worker, making the job more challenging and interesting, but as the machine becomes more sophisticated, as more job skills are built into its workings, a de-skilling trend takes hold. The highly skilled craftsman turns into a moderately skilled or unskilled machine operator. Even Adam Smith understood that machinery, in enhancing labor productivity, would often end up narrowing jobs, turning skilled work into routine work. At worst, he wrote, the factory worker would become "as stupid and ignorant as it is possible for a human creature to become."

That's not the whole picture, of course. In evaluating the long-term effects of automation, we have to look beyond particular job categories. Even as automation reduces the skill requirements of an established occupation, it may contribute to the creation of large new categories of interesting and well-paid work. That's what happened, as the endless-ladder mythologists like to remind us, during the latter stages of the industrial revolution. The efficiencies of assembly lines and other mechanized forms of production pushed down the prices of all sorts of goods, which drove up demand for those goods, which led manufacturers to hire not only lots of blue-collar workers to operate and repair the machines but also lots of white-collar workers to manage the factories, design new products, market and sell the goods, keep the books, and so forth.

The resulting expansion of a consumption-minded, experience-seeking middle class ratcheted up demand for all sorts of other workers, from retail clerks to doctors and nurses to teachers to architects to pilots to journalists to government bureaucrats. A virtuous cycle it most definitely was. What it wasn't was the manifestation of some universal virtuous cycle, some inevitable dynamic in the economy. It was a virtuous cycle very much contingent on its time, and one of the most important of the contingencies was the limited capacity of industrial machinery to take over human work. Even a highly mechanized factory needed lots of people to tend the machines, and most professional and other white-collar jobs lay well beyond the reach of technology.

Times are different now. Machines are different, too. Robots and software programs are still a long way from taking over all human work, but they can take over a lot more of it than factory machines could. It seems pretty clear now that that's one of the main reasons we're seeing persistently depressed demand for workers in many sectors of the economy. What's perhaps less well acknowledged is the spread of the de-skilling phenomenon into so-called knowledge work. As computers become more capable of sensing the environment, performing analyses, and making judgments, they can be programmed to replicate white-collar skills. The remaining professionals and office workers start to look more and more like computer operators, tenders of machines.

There will always be opportunities for individuals to design cool new products, make new scientific discoveries, create new works of art, and think new thoughts. But that says little about the prospects for the labor market in general. There's no guarantee that the deployment of computers is going to open up vast new swathes of interesting, well-paid jobs the way the deployment of factory machines did. Recent experience suggests that workplace computers may have very different consequences. What they seem to be particularly good at is concentrating wealth rather than spreading it, narrowing the work that people do rather than broadening it.

The language that the purveyors of the endless-ladder myth use is revealing. They attribute to technology a beneficent volition. It "frees us" for higher-value tasks and "propels us" into more fulfilling work and "helps us" to expand ourselves. We just need to "allow" the technology to aid us. Much is obscured by such verbs. Technology doesn't free us or propel us or help us. Technology doesn't give a rat's ass about us. It couldn't care less whether we have a great job, a crappy job, or no job at all. It's people who have volition. And the people who design and deploy technologies of production are rarely motivated by a desire to create new jobs or make existing jobs more interesting or expand human potential. They're motivated, as Adam Smith also pointed out, by a desire to make money. Jobs have always been a by-product of the market's invisible hand, not its aim.

The biggest beneficiaries of the endless-ladder myth are those who have gained enormous wealth through the profit-concentrating effects of commercial computers. The myth helps them feel good about themselves. They, after all, are the ones who are setting in motion the virtuous cycle that, in the fullness of time, will bring us all to the nirvana of "higher-and-higher-value work." The myth suits their business interests, too, by conflating those interests with society's interests. Software and robots will solve our problems, if we allow them to.

I'm not saying that it's impossible that we'll soon be blessed with all sorts of great new jobs. The economy's complicated; no one knows what the future's going to bring. I'm saying that we can't take it as a given that that's going to happen, and we certainly shouldn't assume that machines or their owners have the best interests of workers at heart. Ultimately, it's a virtuous cycle—except when it's vicious.

THE LOOM OF THE SELF

April 9, 2014

"IT IS HARD TO RESIST a technology that is also a tool of pleasure," write Sarah Leonard and Kate Losse in the new issue of *Dissent*. "The Luddites smashed their power looms, but who wants to smash Facebook—with all one's photos, birthday greetings, and invitations?"

That's on the money. Things do get messy, *confused*, when the means of production is also the means of communication, the means of expression, the means of entertainment, the means of shopping, the means of fill-in-the-blank. But out of such confusion comes, eventually, simplification, a concentration of effort and effect. Imagine if, at the turn of the nineteenth century, the power loom also served as a social medium. In weaving your quota of cloth, you also wove the story of your life and unfurled it in the public eye. Think of how attached you'd become to your loom. You'd find yourself staying late at the mill, off the clock, working the levers and the foot pedals, the shuttle purring. Hopelessly entangled in the threads, you'd demand a miniature loom that you could use at home, and then an even smaller one that you could carry around with you. Every chance you got, you'd pull out your little loom and start weaving, and all around you others would be doing the same, weaving, weaving, weaving.

I have taken my life from the world, you would say, and I have turned it into cloth, and the pattern in the cloth: that is who I am.

TECHNOLOGY BELOW
AND BEYOND

April 15, 2014

"Neither helplessness nor unbounded enthusiasm and indifference to consequences would have allowed humans to inhabit the earth for very long," observed Bruno Latour, the French sociologist, in a lecture on capitalism and climate change at the Danish Royal Academy earlier this year. "Rather a solid pragmatism, a limited confidence in human cunning, a sane respect for the powers of nature, a great care invested to protect the fragility of human enterprise—these appear to be the virtues for dealing with first nature. Care and caution: a totally mundane grasp of the dangers and of the possibilities of this world of below."

We live in two worlds, according to Latour. There's *first nature*, the earthly "world of below," and there's *second nature*, the transcendent "world of beyond." Second nature reflects our yearning for an existence less fleeting, more constant than the one we lead on earth. Through most of history, second nature manifested itself in myth and religion. Now, argued Latour, it manifests itself in the "laws" of economics. "The world of economy, far from representing a sturdy down-to-earth materialism, a sound appetite for worldly goods and solid matters of fact, is now final and absolute." Purging an economic system of its contingencies and investing it with inexorability tends "to generate for most people who don't benefit from its wealth a feeling of helplessness and for a few people who benefit from it an immense enthusiasm together with a dumbness of the senses." In the face of economic determinism—or maybe we should call it economic eternalism—you get either fatalism or hubris.

What Latour says about our current conception of economics goes equally well for our current conception of technology. Technological progress, too, has come to be viewed as an implacable force beyond our reckoning and control. It generates, to borrow Latour's words, "a prodigious enthusiasm for seizing unbounded opportunities; a dystopian feeling of total helplessness for those who are submitted to its decrees; a complete disinhibition as to the long-term consequences of its action for those who profit from it; a perverse wound of smug superiority in those who have failed to fight its progression." Technology "appears to run more smoothly than nature itself."

Latour finds, in thinking about our shifting sense of economics, a great irony in the "inversion of what is transitory and what is eternal." The irony becomes even stronger when we consider the similar inversion that has taken place in our view of technology. The true glory of technology stems from the possibilities it opens to people in the earthly world of first nature. The glory hinges on technology's contingency, on the way it yields not only to circumstance but to human desire and planning. When technological progress comes to be seen as a transcendent, unyielding force, a force beyond our fashioning, it begins to foreclose opportunities at least as often as it opens them. It starts to hem us in.

"A solid pragmatism, a limited confidence in human cunning, a sane respect for the powers of nature, a great care invested to protect the fragility of human enterprise": would not these earthly virtues serve us equally well in dealing with technology?

OUTSOURCING DAD

June 26, 2014

WHERE, IF ANYWHERE, DO we draw the line between the computer and the self? When, if ever, do we turn to the machine and say, "Back off. I've got this"?

Here's Google's Android chief, Sundar Pichai, offering a peek into the company's vision of our automated future:

> Today, computing mainly automates things for you, but when we connect all these things, you can truly start assisting people in a more meaningful way. . . . If I go and pick up my kids, it would be good for my car to be aware that my kids have entered the car and change the music to something that's appropriate for them.

What's revealing is not the triviality of the scenario—that billions of dollars might be invested in developing a system that senses when your kids get in your car and then seamlessly cues up "Baby Beluga"— but what the urge to automate small acts of affiliation and affection says about Pichai and his colleagues. With this offhand example, Pichai gives voice to Silicon Valley's reigning assumption, which can be boiled down to this: Anything that can be automated should be automated. If it's possible to program a computer to do something a person can do, the computer should do it.

Missing from this view is any consideration of the pleasures and responsibilities of everyday life. Pichai doesn't seem to have entertained the possibility that much of the joy of parenting lies in the little, inconsequential gestures that parents make on behalf of or in concert with their kids, like picking out a song to play in the car.

TAKING MEASUREMENT'S MEASURE

August 26, 2014

"What can't be measured can't be managed," goes the old saw. But what the business theorist Peter Drucker is reported to have actually said was "What gets measured gets managed," which is altogether different and altogether wiser. The wisdom becomes clearer when we get the rest of Drucker's remark: "What gets measured gets managed—even when it's pointless to measure and manage it, and even if it harms the purpose of the organization to do so."

It's dubious and dangerous, Drucker is saying, to mistake what's measurable for what's important. But he's also saying something much more radical, even subversive: Some things that can be measured shouldn't be.

A whole counterculture could, in our big-data moment, be constructed on that one thought. Can you imagine Google or Amazon or Facebook announcing, "We have decided to stop measuring stuff in order to spend some time considering what's actually worth measuring"? No, today's ethos is simpler, easier to execute: "If you measure it, the meaning will come."

"Measure" itself has a few meanings, and it's worth keeping them all in mind. Speaking to college students in 1956, Robert Frost said, "I am always pleased when I see someone making motions like this—like a metronome. Seeing the music measured. Measure always reassures me. Measure in love, in government, measure in selfishness, measure in unselfishness." Measure in measurement, too, would seem advisable.

SMARTPHONES ARE HOT

October 21, 2014

THE LIGHTBULB, MARSHALL MCLUHAN wrote, is an example of a medium without content. Walk into a dark room and hit the light switch, and the bulb generates a new environment even though the bulb transmits no information. The idea of a medium without content is hard to grasp. It doesn't make sense in the context of our assumptions about media. But it's fundamental to understanding McLuhan's contention that the medium is the message—that every medium creates an environment independent of the content or information it transmits.

So what are we to make of the smartphone, the medium of the moment, our portable environment? If, as McLuhan argued, the content of any new medium is an old medium, the content of the smartphone would seem to be all media: telephone, television, radio, cinema, printed book, electronic book, comic book, record, MP3, newspaper, magazine, letter, newsletter, email, peep show, library, school, lecture, ATM, desktop, laptop, love note, medical record, rap sheet. Contentwise, the smartphone is Whitmanesque: it contains multitudes. The smartphone is what happens when the architecture of media collapses. It's a black hole full of light: information supercompressed but radiant. In its singularity, it might be described as the first postmedia medium. Its circuitry dissolves plurality; the media becomes the medium.

Bursting with information, the smartphone is, in McLuhan's terms, a hot medium, maybe the hottest imaginable. It invades the sensorium of its user with an absolute imperialist zeal. Flooding the visual sense, it allows no signal but its own. To look into the screen of a smartphone is to be lost to the world. Like every hot medium, the smartphone isolates and fragments the self. It individualizes, alienates. Not only does it reverse what McLuhan described as the coolness of the aural tele-phone, turning it into a superheated visual medium, but it reverses the entire retribalization pattern that McLuhan saw emerging from electric media. The smartphone out-detribalizes even the printed book. The smartphone's "interactivity" is a ruse, for the only activity it allows is the activity it mediates. Its psychic dominance precludes involve-ment and participation.

But that can't be right. What does one do with a smartphone but participate—interact, converse, communicate, shop, create, get involved? Here we find the conundrum of the smartphone, the conun-drum of our new artificial environment—and the conundrum that wraps around McLuhan's hot/cool media dialectic.

In a 1967 essay, the critic Richard Kostelanetz wrote that McLuhan's books "offer a cool experience in a hot medium." The low-definition ambiguity of the writing fights against the high-definition clarity of the printed word; the information demands the reader's involvement while the medium forbids it. It may be that the smartphone is of a similar nature, hot and cool at once (but never lukewarm). At the very least, one could say that the smartphone creates an environment that encourages participation at a distance: participation as performance.

The smartphone retribalizes by putting us always on display, by eat-ing away at our sense of the private self, but it detribalizes by isolat-ing us in an abstract world, a world of our own. You hit the switch, and the light comes on and you find yourself in an empty room full of people. To put it another way: Participation is the content of the smartphone, and the content, as McLuhan wrote, is "the juicy piece of meat carried by the burglar to distract the watchdog of the mind." The illusion of involvement conceals its absence. Here comes Walt Whit-man, alone and alienated, dreaming dreams of connection, turning a barbaric yawp into silent words on a flat page.

DESPERATE SCRAPBOOKERS

November 7, 2014

TOWARD THE END OF the nineteenth century, a Wyoming prostitute named Monte Grover began cutting snippets of poetry out of newspapers and magazines and pasting them into a scrapbook. She used the clippings to "construct an idealized life by isolating a set of values that she found around her," write the authors of *The Scrapbook in American Life.* Grover's scrapbook contained "her inner identity and her best self."

So it goes:

> People today, more than a hundred years later, find their identities recorded and inscribed in bureaucratic files and data banks; their official human identities are found in X rays, birth certificates, driver's licenses, and DNA samples. But a scrapbook represents a construction of identity outside these formalized and authoritative records. It is the self that guides the scissors and assembles the scraps.

It struck me, as I was scrolling through some guy's Tumblr yesterday, that the scrapbook has become our essential cultural form, the artifact that defines the time. Watching TV shows and films, reading books and articles, listening to songs: they all still have their places in our lives, sure. But it's scrapbooking, particularly of the unbound, online variety, that consumes us. If we're not arranging our own scraps, we're rummaging through the scraps of others.

A scrapbooking metaphor—"cut and paste"—has long suffused our experience of computers. Now, the scrapbook *is* the user interface. The cloud is our great shared scrapbook in the sky.

Pinterest makes its scrapbooky nature most explicit, but, really, all social networks are scrapbooks: Facebook, Twitter, Tumblr, Instagram, YouTube, LinkedIn. Even the more basic messaging systems—email, texting—feel more and more scrappy, now that we don't bother to delete the messages. ("It deepens like a coastal shelf," wrote Philip Larkin, and indeed it does.) Blogs are scrapbooks. Huffington Post and Medium are scrapbooks. A tap of a Like button is nothing if not a quick scissoring.

Scrapbooking and data-mining are the yang and the yin of the web: light and dark, aboveground and belowground, exposed and secret. Today's scrapbooks serve both as a counterweight to the bureaucratic file and as part of the file's contents. The Eloi's pastime is fodder for the Morlocks.

Inherently retrospective—a means of preemptively packaging the present as memory—scrapbooking is a melancholy art. Pressed insistently forward, we spend our time arranging the bits and pieces of our lives into something we think looks something like us. If the scrapbook of old was familial and semiprivate, the new one is social and relentlessly public. It's still a melancholy form, but now it's an anxious one, too. It's one thing to construct an idealized life, a "best self," for your own perusal; it's another thing to construct one for all to see.

"It appears, then, that scrapbook-making as a ritualized, order-inducing gesture is both an acknowledgement of and a response to the heightened sense of fragmentation which has attended the experience of modernity," wrote Tamar Katriel and Thomas Farrell in their 1991 article "Scrapbooks as Cultural Texts." They may be right. And maybe the appeal of the digital form of scrapbooking is that it's all-encompassing and never-ending. As long as you're arranging your fragments, you don't have time to realize that they're fragments. The lack of coherence just means that a piece is still missing.

OUT OF CONTROL

January 17, 2015

Machines that think think like machines. That may disappoint those who look forward, with dread or longing, to a robot uprising. For most of us, it's reassuring. Our thinking machines aren't about to leap beyond us intellectually, much less turn us into their servants or pets. They're going to continue to do the bidding of their human programmers.

Much of the power of artificial intelligence stems from its very mindlessness. Immune to the vagaries and biases that attend conscious thought, computers can perform their lightning-quick calculations without distraction or fatigue, doubt or emotion. The coldness of their thinking complements the heat of our own.

Things get sticky when we start looking to computers to perform not as our aides but as our replacements. That's what's happening now, and quickly. Thanks to advances in artificial-intelligence routines, today's thinking machines can sense their surroundings, learn from experience, and make decisions autonomously, often at a speed and with a precision that are beyond our own ability to comprehend, much less match. When allowed to act on their own in a complex world, whether embodied as robots or simply outputting algorithmically derived judgments, mindless machines carry enormous risks along with their enormous powers. Unable to question their own actions or appreciate the consequences of their programming—unable to understand the context in which they operate—they can wreak havoc, either as a result of flaws in their programming or through the deliberate aims of their programmers.

We got a preview of the dangers of autonomous software on the morning of August 1, 2012, when Wall Street's biggest trading outfit,

Knight Capital, switched on a new, automated program for buying and selling shares. The software had a bug hidden in its code, and it immediately flooded exchanges with irrational orders. Forty-five minutes passed before Knight's programmers were able to find and fix the problem. Forty-five minutes isn't long in human time, but it's an eternity in computer time. Oblivious to its errors, the software made more than four million deals, racking up $7 billion in errant trades and nearly bankrupting the company. Yes, we know how to make machines think. What we don't know is how to make them thoughtful.

All that was lost in the Knight fiasco was money. As software takes command of more economic, social, and military processes, the costs of glitches, breakdowns, and unforeseen effects will only grow. Compounding the danger is the invisibility of software code. As individuals and as a society, we increasingly depend on programmed routines that we neither see nor understand. Their workings, and the motivations and intentions that shape their workings, are hidden from us. That creates an imbalance of power, and it leaves us open to clandestine surveillance and manipulation. Last year we got some hints about the ways that social networks conduct secret psychological tests on their members through the manipulation of information feeds. As computers become more adept at monitoring us and shaping what we see and do, the potential for abuse grows.

During the nineteenth century, society faced what the historian James Beniger described as a "crisis of control." The technologies for processing matter had outstripped the technologies for processing information, and people's ability to monitor and regulate industrial and related processes had in turn broken down. The control crisis, which manifested itself in everything from train crashes to supply-and-demand imbalances to interruptions in the delivery of government services, was eventually resolved through the invention of systems for automated data processing, such as the punch-card tabulator that Herman Hollerith built for the U.S. Census Bureau. Information technology caught up with industrial technology, enabling people to bring back into focus a world that had gone blurry.

Today, we face another control crisis, though it's the mirror image of the earlier one. We're struggling to bring under control the very thing that helped us reassert control at the start of the twentieth century: information technology. Our ability to gather and process data, to manipulate information in all its forms, has outstripped our ability to monitor and regulate data processing in a way that suits our societal and personal interests. Resolving this new control crisis will be one of the great challenges in the years ahead. The first step in meeting the challenge is to recognize that the risks of artificial intelligence don't lie in some dystopian future. They are here now.

OUR ALGORITHMS, OURSELVES

March 8, 2015

PUT YOURSELF IN THE shoes of Mario Costeja González. In 1998, the Spaniard found himself in a financial bind. He had defaulted on a debt and to pay it off was forced to sell some real estate at auction. The sale was duly noted in the venerable Barcelona newspaper *La Vanguardia*. The matter settled, Costeja went on with his life as a legal graphologist, an interpreter of handwriting. The default and the auction, as well as the thirty-six-word press notice about them, faded from public memory.

Nearly a dozen years later, in 2009, the episode sprang back to life. *La Vanguardia* put its archives online, Google's web-crawling "bot" sniffed out the old article about the auction, the article was automatically added to the search engine's database, and a link to it began popping into prominent view whenever someone in Spain did a search on Costeja's name. Costeja was dismayed. It seemed unfair to have his reputation sullied by an out-of-context report on an old personal problem that had long ago been resolved. Presented without explanation in search results, the article made him look like a deadbeat. He felt his dignity was at stake.

Costeja lodged a formal complaint with the Spanish government's data-protection agency. He asked the regulators to order *La Vanguardia* to remove the article from its website and to order Google to stop linking to the notice in its search results. The agency refused to act on the newspaper request, citing the legality of the article's original publication, but it agreed with Costeja about the unfairness of the Google listing. It told the company to remove the auction story from its results. Appalled, Google appealed the decision, arguing that in listing the story it was merely highlighting information published elsewhere.

The dispute made its way to the Court of Justice of the European Union in Luxembourg, where it became known as the "right to be forgotten" case. On May 13 last year, the high court issued its final, binding decision. Siding with Costeja and the Spanish data-protection agency, the justices ruled that Google was obligated to obey the order and remove the *La Vanguardia* piece from its search results. Suddenly, European citizens had the right to get certain unflattering information about them deleted from search engines.

Most Americans, and quite a few Europeans, were flabbergasted by the decision. They saw it not only as unworkable (how can a global search engine processing some six billion searches a day be expected to evaluate the personal grouses of individuals?), but also as a threat to the free flow of information online. Many accused the court of licensing censorship or even of creating "memory holes" in history. Some worried that the ruling could "break the internet."

The heated reactions, however understandable, were off the mark. They reflected a misreading of the decision. The court had not established a "right to be forgotten." That essentially metaphorical phrase is mentioned only in passing in the ruling, and its attachment to the case has proven a distraction. In an open society, where freedom of thought and speech are protected, where people's thoughts and words are their own, a right to be forgotten is as untenable as a right to be remembered. What the case was really about was an individual's right not to be systematically misrepresented. But even putting the decision into those more modest terms is misleading. It implies that the court's ruling was broader than it actually was.

The essential issue the justices were called upon to address was how, if at all, a 1995 European Union policy on the processing of personal data, the so-called Data Protection Directive, applied to companies that, like Google, engage in the large-scale collection and dissemination of information online. The directive had been enacted to ease the cross-border exchange of data while also establishing privacy and other protections for citizens. "Whereas data-processing systems are designed to serve man," the policy reads, "they must, whatever the nationality or residence of natural persons, respect their fundamental rights and freedoms, notably the right to privacy, and contribute to economic and social progress, trade expansion and the well-being of individuals."

To shield people from abusive or unjust treatment, the directive imposed strict regulations on businesses and other organizations that act as "controllers" of the processing of personal information. It required, among other things, that any data distributed by such controllers be not only accurate and up-to-date, but fair, relevant, and "not excessive in relation to the purposes for which they are collected and/or further processed." What the directive left unclear was whether companies that aggregated information produced by others—the Googles and Facebooks of the world—fell into the category of controllers. That was what the court had to decide.

Search engines, social networks, and other online data aggregators have always presented themselves as playing a neutral and essentially passive role when it comes to the processing of information. They're not creating the content they distribute—that's done by publishers in the case of search engines, or by individual members in the case of social networks. Rather, they're simply gathering the information and arranging it in a useful form. This view, tirelessly promoted by Google—and used by the company as a defense in the Costeja case—has been embraced by much of the public. It has become the default view. When Wikipedia cofounder Jimmy Wales, in criticizing the European court's decision, said, "Google just helps us to find the things that are online," he was not only mouthing the company line, he was expressing the popular conception of internet businesses.

The court took a different view. Online aggregation is not a neutral act, it ruled, but a transformative one. In collecting, organizing, and ranking information, a search engine is creating something new, a product that reflects the business's own editorial intentions and judgments, as expressed through its information-processing algorithms. "The processing of personal data carried out in the context of the activity of a search engine can be distinguished from and is additional to that carried out by publishers of websites," the justices wrote. Since the presentation of search results influence "the fundamental rights to privacy and to the protection of personal data, the operator of the search engine . . . must ensure, within the framework of its responsibilities, powers and capabilities, that the activity meets the requirements of [the Data Protection Directive] in order that the guarantees laid down by the directive may have full effect."

The European court did not pass judgment on the guarantees established by the Data Protection Directive, nor on any other existing or prospective laws or policies pertaining to the processing of personal information. It did not tell society how to assess or regulate the activities of aggregators. It did not even offer an opinion as to the process companies or lawmakers should use in deciding which personal information warranted exclusion from search results—an undertaking every bit as thorny as it's been made out to be. What the justices did, with perspicuity and prudence, was provide us with a way to think rationally about the algorithmic manipulation of digital information and the social responsibilities it entails. The interests of a powerful international company like Google, a company that provides an indispensable service to many people, do not automatically trump the interests of a lone individual. When it comes to the operation of search engines and other information aggregators, fairness is at least as important as expedience.

We have had a hard time thinking clearly about companies like Google and Facebook because we have never before had to deal with companies like Google and Facebook. They are something new in the world, and they don't fit neatly into our legal and cultural templates. Because they operate at an unimaginable scale and speed, carrying out millions of informational transactions every second, we've tended to see them as vast, faceless, dispassionate computers—as information-processing machines that exist outside the realm of human intention and control. That's a misperception. Modern computers and computer networks enable human judgment to be automated, but it's still human judgment. Algorithms are constructed by people, and they reflect the interests, biases, and flaws of their makers. As a society, we have an obligation to carefully examine and, when appropriate, judiciously regulate those algorithms.

Ten months have passed since the court's ruling, and we now know that it is not going to break the internet. The web still works. Google has a process in place for adjudicating requests for the removal of personal information—it accepts about 40 percent of them—just as it has a process in place for adjudicating requests to remove material under copyright. Last month, Google's Advisory Council on the Right to Be Forgotten issued a report that put the ruling and the company's response into context. "In fact," the council wrote, "the Ruling does not establish a general Right to Be Forgotten. Implementation of the Ruling does not have the effect of 'forgetting' information about a data subject. Instead, it requires Google to remove links returned in search results based on an individual's name when those results are 'inadequate, irrelevant or no longer relevant, or excessive.' Google is not required to remove those results if there is an overriding public interest in them 'for particular reasons, such as the role played by the data subject in public life.'" It is possible, the case of Mario Costeja González tells us, to strike a reasonable balance among an individual's interests, the interests of the public in finding information quickly, and the commercial interests of internet companies.

TWILIGHT OF THE IDYLLS

March 31, 2015

THE SILICON VALLEY GUYS have a new hobby: driving fast cars around private tracks. They love it. "When you're really in the zone in a racecar, it's almost meditative," Google executive Jeff Huber tells Farhad Manjoo of the *Times*. Adds Yahoo senior vice president Jeff Bonforte, "Your brain is so happy that it washes over you." The Valley guys are a little nervous about the optics of their pastime—"Try to tone down the rich guy hobby thing," angel investor and ex-Googler Joshua Schachter instructs Manjoo—but the "visceral thrill" of driving has nevertheless made it "the Valley's 'it' hobby."

The Valley guys are rushing to rent out racetracks and strap themselves into Ferraris at the very moment that they're telling the rest of us how miserable driving is, and how happy we'll all feel when robots take the wheel. Jazzed by a Googler's TED Talk on driverless cars, MIT automation expert Andrew McAfee says that the Googlemobile will "free us from a largely tedious task." Writes *Wired* transport reporter Alex Davies, "Liberated from the need to keep our hands on the wheel and eyes on the road, drivers will become riders with more time for working, leisure, and staying in touch with loved ones"—in other words, they'll be able to spend more time on their phones. When Astro Teller, head of the Google X lab, watches people drive by in their cars, all he hears is a giant sucking sound, as potentially productive minutes pour down the drain of a vast time sink: "There's over a trillion dollars of wasted time per year we could collectively get back if we didn't have to pay attention while the car took us from one place to another."

Driving on a private track may be pleasantly meditative, even joy-inducing, but driving on public thoroughfares is just a drag.

What's curious here is that the descriptions of everyday driving offered with such certainty by the champions of driverlessness are at odds with what we know about people's attitudes toward and experience of driving. People like to drive. Surveys and other research consistently show that most of us enjoy being behind the wheel. We find driving relaxing and fun and even, yes, liberating—a respite from the demands of our workaday lives. Seeing driving as a "problem" because it prevents us from being productive gets the story backward. What's freeing about driving is the very fact that it gives us a break from the pressure to be productive. Driving allows us to take pleasure from simply *doing.*

That doesn't mean we're blind to automotive miseries. When researchers talk to people about driving, they hear plenty of complaints about traffic jams and grinding commutes and bad roads and parking hassles and all the rest. Our attitudes toward driving are complicated, but on balance we like to have our hands on the wheel and our eyes on the road, not to mention our foot on the gas. About 70 percent of Americans say they "like to drive," while only about 30 percent consider it "a chore," according to a 2006 Pew survey. A survey of millennials, released earlier this year by MTV, found that, contrary to common wisdom, most young people enjoy cars and driving, too. Seventy percent of Americans between the ages of 18 and 34 say they like to drive, 72 percent say they'd rather give up texting for a week than give up their car for the same period, and 85 percent look forward to one day owning their "dream car."

The percentage of people who like to drive has fallen a bit in recent years as traffic has worsened—80 percent said they liked to drive in a 1991 Pew survey—but it's still sky high, and it belies the dreary picture of driving painted by Silicon Valley. You don't have to be wealthy enough to buy a Porsche or rent out a racetrack to enjoy the small pleasures of driving. They can be felt on the open road as well as the closed track.

In suggesting that driving is no more than a boring, productivity-sapping waste of time, the Valley guys are mistaking a personal bias for a universal truth. And they're blinding themselves to the social and cultural challenges they're going to face as they try to convince people to be passengers rather than drivers. Even if all the technical hurdles to achieving perfect vehicular automation are overcome—and despite rosy predictions, that remains a sizable *if*—the developers of autonomous cars are going to discover that the psychology of driving is far more complicated than they assume and far different from the psychology of being a passenger. Back in the 1970s, the public rebelled, en masse, when the federal government, for seemingly solid safety and fuel-economy reasons, imposed a national fifty-five-mile-per-hour speed limit. The limit was repealed. If you think everyone's going to happily hand the keys over to a robot, you're crazy.

Silicon Valley seems to have trouble appreciating, or even understanding, what I'll term "informal experience." It's only when driving is formalized—removed from everyday life, transferred to a specialized facility, performed under a strict set of rules, and understood as a self-contained recreational event—that it can be conceived of as being pleasurable. When it's not a recreational routine, when it's performed out in the world, as part of everyday life, then driving, in the Valley view, can only be understood within the context of another formalized realm of experience: that of productive busyness. Every experience has to be cleanly defined, has to be categorized. There's a place and a time for recreation, and there's a place and a time for production.

This discomfort with the informal, with experience that is psychologically unbounded, that flits between and beyond categories, can be felt in a lot of the Valley's consumer goods and services. Many personal apps and gadgets have the effect, or at least the intended effect, of formalizing informal activities. Once you strap on a Fitbit, you transform what might have been a pleasant walk in the park into a program of physical therapy. A passing observation that once might have earned a few fleeting smiles or shrugs before disappearing into the ether is now, thanks to the distribution systems of Facebook and Twitter, encapsulated as a product and subjected to formal measurement; every remark gets its own Nielsen rating.

What's the source of this crabbed view of experience? I'm not sure. It may be an expression of a certain personality type. It may be a sign of the market's continuing colonization of the quotidian. I'd guess it also has something to do with the rigorously formal qualities of programming itself. The universality of the digital computer ends—comes to a crashing halt, in fact—where informality begins.

THE ILLUSION
OF KNOWLEDGE

April 2, 2015

THE INTERNET MAY BE making us shallow, but it's making us think we're deep.

A new Yale study reveals that searching the web provides people with an "illusion of knowledge." They start to confuse what's online with what's in their head, which gives them an exaggerated sense of their own smarts. The researchers divided test subjects into two groups. One group spent time searching the web, the other stayed offline, and then both groups estimated, in a variety of ways, their understanding of various topics. The experiments consistently showed that searching the web deludes people into believing they know more than they really do. The effect held, the researchers report, "even after controlling for time, content, and features of the search process."

The effect isn't limited to the particular subjects people explore online. It's more general than that. Doing searches on one topic inflates people's sense of how well they understand other, unrelated topics. To make sure that the searchers' overconfidence stemmed from a misperception of the depth of knowledge in their own heads (rather than reflecting a trust in Google's ability to deliver relevant knowledge), the psychologists, in one of the experiments, had the test subjects make estimates of their brain activity. Those who had been searching the net before the task rated their brain activity as being significantly stronger than did the control group that hadn't been looking up information online.

Similar misperceptions may be produced by consulting other sources of information, the researchers note, but the illusion is probably much stronger with the web, given its unprecedented scope. Unlike earlier stores of knowledge such as books and libraries, "the Internet is nearly always accessible, can be searched efficiently, and provides immediate feedback." As a result, the net likely becomes more "integrated with the human mind" and promotes "much stronger illusions of knowledge."

This is just one study, but it's consistent with other research into the web's influence on how our minds construct, or don't construct, personal knowledge. The University of Colorado's Adrian Ward has found evidence that the shift from "biological information storage" toward "digital information storage" may "have large-scale and long-term effects" on the way we think. In "How Google Is Changing Your Brain," a 2013 *Scientific American* article written with the late memory expert Daniel Wegner, Ward reported on experiments that, like the Yale work, reveal how web searching "gives people the sense that the Internet has become part of their own cognitive tool set." People "take credit for knowing things that were a product of Google's search algorithms." There's an irony here, Ward and Wegner noted: "The advent of the 'information age' seems to have created a generation of people who feel they know more than ever before—when their reliance on the Internet means that they may know ever less about the world around them." Ignorance is bliss, particularly when it's mistaken for knowledge.

WIND-FUCKING

May 15, 2015

VOCABULARY IS RARELY SO lush, so dense with branch and twig, as in the realm of flora and fauna. Plants and animals go by all sorts of strange and evocative names depending on where you are and whom you're talking with. One local term for the kestrel, reports Robert Macfarlane in an article in *Orion*, is "wind-fucker." Having learned the word, he writes, "it is hard now not to see in the pose of the hovering kestrel a certain lustful quiver."

I'm reminded of Seamus Heaney's translation of a Middle English poem, "The Names of the Hare":

> *Beat-the-pad, white-face,*
> *funk-the-ditch, shit-ass.*

It goes on like that for a few dozen lines, and each name in the litany brings you a little closer to the nature of the beastie.

Macfarlane reports that the *Oxford Junior Dictionary* is being pruned of words describing the stuff of the natural world. They're being replaced by words denoting the abstractions and symbols of the digital and bureaucratic spheres.

Oxford University Press revealed a list of the entries it no longer felt to be relevant to a modern-day childhood. The deletions included acorn, adder, ash, beech, bluebell, buttercup, catkin, conker, cowslip, cygnet, dandelion, fern, hazel, heather, heron, ivy, kingfisher, lark, mistletoe, nectar, newt, otter, pasture, and willow. The words introduced to the new edition included attachment, block-graph, blog, broadband, bullet-point, celebrity, chatroom, committee, cut-and-paste, MP3 player, and voice-mail.

They yanked out *bluebell* and stuck in *bullet-point*? What shit-asses. "The substitutions made in the dictionary—the outdoor and the natural being displaced by the indoor and the virtual—are a small but significant symptom of the simulated life we increasingly live," writes Macfarlane. "A basic literacy of landscape is falling away up and down the ages." The changes in the dictionary don't just testify to our weakening grasp on nature. Something else is being lost: "a kind of word magic, the power that certain terms possess to enchant our relations with nature and place."

I'm sure that many will label Macfarlane a romantic. I've begun to notice that *romantic* is replacing *Luddite* and *nostalgist* as the insult-of-choice deployed by technophiles to dismiss anyone with more expansive interests than their own. That, too, is telling. It has always been a sin against progress to look backward. Now it is also a sin against progress to look inward. And so, fading from sight and imagination alike, the world becomes ever vaguer to us—not mysterious but peripheral, its things unworthy even of being named. Who now would think of the wind as something that might be fucked?

THE SECONDS ARE
JUST PACKED

June 9, 2015

"Everything is going too fast and not fast enough," laments Warren Oates, playing a decaying gearhead called G.T.O., in Monte Hellman's 1971 masterpiece *Two-Lane Blacktop*. I can relate. The faster the clock spins, the more I feel as if I'm stuck in a slo-mo GIF loop.

It's weird. We humans have been shown to have remarkably accurate internal clocks. Take away our wristwatches and our cell phones, dim the LEDs on all our appliances, and we can still make pretty good estimates about the passage of minutes and hours. Our brains have adapted well to mechanical time-keeping devices. But our time-tracking faculty goes out of whack easily. Our perception of time is subjective; it changes with our circumstances. When things are happening quickly around us, delays that would otherwise seem brief begin to feel interminable. Seconds stretch out. Minutes go on forever. "Our sense of time," observed William James in *Principles of Psychology*, "seems subject to the law of contrast."

In a 2009 article in the *Philosophical Transactions of the Royal Society*, French psychologists Sylvie Droit-Volet and Sandrine Gil described what they call the paradox of time. "Although humans are able to accurately estimate time as if they possess a specific mechanism that allows them to measure time," they wrote, "their representations of time are easily distorted by the context." They describe how our sense of time changes with our emotional state. When we're agitated or anxious, for instance, time seems to crawl; we lose patience. Our social milieu, too, influences the way we experience time. Studies show, write Droit-Volet and Gil, "that individuals match their time with that of others." The "activity rhythm" of those around us alters our own perception of the passing of time.

"A compression of time characterizes the life of the century now closing," wrote James Gleick in his 1999 book *Faster*. But that was nothing new. Such compression characterized, as well, the preceding century. "The dreamy quiet old days are over and gone forever," lamented a cultural critic named William Smith in 1886, "for men now live, think and work at express speed." I suspect it would take no more than a minute or two of Googling to discover a quotation from one of the ancients bemoaning the horrific speed of contemporary life. The past has always had the advantage of seeming, and probably being, less hurried than the present.

Still, something has changed in the last few years. Given what we know about the variability of our time sense, it seems clear that information and communication technologies would have a particularly strong effect on our perception of time. After all, they often determine the pace of the events we experience, the speed with which we're presented with new information and stimuli, and even the way we converse with others. That's been true for a long time—the newspaper, the telephone, and the television all quickened the speed of life—but the influence must be all the stronger now that we carry powerful and extraordinarily fast computers around with us all day long. Our gadgets train us to expect near-instantaneous responses to our actions, and we quickly get frustrated and annoyed at even brief delays.

I know from my own experience with computers that my perception of time has been changed by technology. If I go from using a fast computer or web connection to using even a slightly slower one, processes that take just a few seconds longer—waking the machine from sleep, launching an application, opening a web page—seem almost intolerably slow. Never before have I been so aware of, and annoyed by, the passage of mere seconds.

Research on web users makes it clear that this is a general phenomenon. Back in 2006, a famous study of online retailing found that a large percentage of shoppers would abandon a merchant's site if its pages took four seconds or longer to load. In the few years since, the so-called Four-Second Rule has been repealed and replaced by the Quarter-of-a-Second Rule. Studies by companies like Google and Microsoft now find that it takes a delay of just 250 milliseconds in page loading for people to start abandoning a site. "Two hundred fifty milliseconds, either slower or faster, is close to the magic number now for competitive advantage on the web," a top Microsoft engineer said in 2012. To put that into perspective, it takes about the same amount of time for you to blink your eyes.

A recent study of online video viewing provides more evidence of the way advances in media and networking technology reduce the patience of human beings. The researchers, affiliated with the networking firm Akamai Technologies, studied a huge database that documented twenty-three million video views by nearly seven million people. They found that people start abandoning a video in droves after a two-second delay. That won't come as a surprise to anyone who has had to wait for a YouTube clip to begin after clicking the Start button. More interesting is the study's finding of a causal link between higher connection speeds and higher abandonment rates. Every time a network gets quicker, we get antsier. "Every millisecond matters," says a Google engineer.

As we experience faster flows of information online, we become, in other words, less patient people. But impatience is not just a network effect. The phenomenon is amplified by the constant buzz of Facebook, Twitter, Snapchat, texting, and social networking in general. Society's "activity rhythm" has never been so harried. Impatience is a contagion spread from gadget to gadget.

All of this has obvious importance to anyone involved in producing online media or running data centers. But it also has implications for how all of us think, socialize, and in general live. If we assume that networks will continue to get faster—a pretty safe bet—then we can also conclude that we'll become more and more impatient, more and more intolerant of even the slightest delay between action and response, between desire and its fulfillment. As a result, we'll be less likely to experience anything that requires waiting, that doesn't provide instant gratification. That has cultural as well as personal consequences. The greatest of works—in art, science, politics, whatever—tend to take time and patience both to create and to appreciate. The deepest experiences can't be measured in fractions of seconds.

It's not clear whether a technology-induced loss of patience persists even when we're not using the technology. But I would hypothesize (from what I see in myself and others) that our sense of time is indeed changing in a lasting way. Digital technologies are training us to be more conscious of and more antagonistic toward delays of all sorts— and perhaps more intolerant of moments of time that pass without the arrival of new messages or other stimuli. Call it the patience deficit. Because our experience of time is so important to our experience of life, these kinds of media-induced changes in our perceptions can have particularly broad consequences. How long are you willing to wait for the next new thing to arrive? How many empty seconds can you endure?

MUSIC IS THE UNIVERSAL LUBRICANT

June 23, 2015

IN ANNOUNCING A FREE VERSION of its music streaming service today, Google also revealed something interesting about its conception of music:

> At any moment in your day, Google Play Music has whatever you need music for—from working, to working out, to working it on the dance floor—and gives you curated radio stations to make whatever you're doing better. Our team of music experts, including the folks who created Songza, crafts each station song by song so you don't have to.

This marks a continuation of Google's promotion of what it terms "activity-based" music. Last year, soon after it acquired Songza, a company that specializes in "curating" playlists to suit particular moods and activities, Google rejiggered its music service to emphasize its practicality:

> If you're a Google Play Music subscriber, next time you open the app you'll be prompted to play music for a time of day, mood or activity. Choose an activity to get options for several music stations to make whatever you're doing even better—whether it's a station for a morning workout, songs to relieve stress during traffic, or the right mix for cooking with friends. Each station has been hand-crafted—song by song—by our team of music experts (dozens of DJs, musicians, music critics and ethnomusicologists) to give you the exact right song for the moment.

This is the democratization of the Muzak philosophy. Music becomes an input, a factor of production. Listening to music is not itself an "activity"—music isn't an end in itself—but rather an enhancer of other activities, each of which must be clearly demarcated. Here's a sampling of the discrete activities—"jobs" might be the more accurate term—that Google lets you choose from in ordering up units of music:

Barbecuing
Being Romantic
Breaking Up
Coding
Cooking
Daydreaming
Drinking
Driving
Entering Beast Mode
Falling in Love
Family Time
Getting Cozy
Getting Married
Girls Night Out
Having Friends Over
Having Fun at Work
Pregaming
Raising Your Kids
Relaxing
Studying
Waking Up
Working
Working Out
Yoga

In a way, this seems inevitable. Popular music today is created by teams of experts trained in digital production techniques. Working with raw material supplied by a songwriter, specialists in beats, synthetic instrumentation, lyrics, hooks, melody, and vocal manipulation collaborate to assemble hits perfectly tuned to the ear of the mass market. The manufacture of pop songs has been so thoroughly industrialized that it makes the old Motown "hit factory" look like a sewing circle. Google is doing to listening what's already been done to recording.

Once you accept that music is an input, a factor of production, you'll naturally seek to minimize the cost and effort required to acquire the input. And since music is "context" rather than "core," to borrow Geoffrey Moore's famous categorization of business inputs, simple economics dictates that you outsource the supply of music rather than invest personal resources—time, money, attention, passion—in supplying it yourself. You should, as Google suggests, look to professionals to "craft" your musical inputs "song by song" so "you don't have to." To choose your own songs, or even to develop the personal taste in music required to choose your own songs, would be wasted labor, a distraction from the series of essential jobs that give structure and value to your days.

Art is an industrial lubricant. By removing the friction from everyday activities, it makes for more productive lives.

TOWARD A UNIFIED
THEORY OF LOVE

July 30, 2015

WE LIVE MYTHICALLY, EVEN the most calculating of us. In the middle of a bromidic Q&A session on Facebook last month, Mark Zuckerberg fielded a question from the cosmologist Stephen Hawking: "I would like to know a unified theory of gravity and the other forces. Which of the big questions in science would you like to know the answer to and why?" Zuckerberg replied that he was "most interested in questions about people," and he gave some examples of the questions about people that he found most interesting. "What will enable us to live forever?" was one. "How can we empower humans to learn a million times more?" was another.

He then divulged something interesting, if not unexpected, about his perception of the social world: "I'm also curious about whether there is a fundamental mathematical law underlying human social relationships that governs the balance of who and what we all care about. I bet there is." Call it the unified theory of love.

Zuckerberg's answer underscores, yet again, what an odd choice we made when we picked a person to oversee the world's chief social network. We've placed our social lives in the hands of a maladroit young man who believes that human relations and affiliations can be reduced to equations.

> *The fault, dear Brutus, is not in our stars,*
> *But in ourselves, that we are underlings.*

What Brutus saw in stars, Zuckerberg sees in data. Both believe that human affairs are governed by fate.

It's not hard to understand the source of Zuckerberg's misperception. Human beings, like ants or chickens, share a certain bundle of tendencies, a certain nature, and if you analyze our behavior statistically, that nature will evidence itself in mathematical regularities. Facebook's founder is hardly the first to confuse the measurement of a phenomenon with the cause of the phenomenon. If some amount of data reveals a pattern, then, surely, more data will reveal "a fundamental mathematical law."

Zuckerberg's belief that social relations are the output of a cosmic computer running a cosmic algorithm is more than just the self-serving fantasy of a man who has made a fortune by seeing people as nodes in a mathematical graph. It's an expression, however extreme, of a new form of behavioralism that has recently come into vogue, pulled along in the slipstream of the excitement over "big data."

From the mid-1950s to the mid-1960s, sociological thinking in the United States was dominated by the behavioralist school. Heirs of the earlier positivists, behavioralists believed that social structures and dynamics could be understood only through the rigorous, scientific analysis of hard data. David Easton, a prominent University of Chicago political scientist, laid out the tenets of behavioralism in a 1962 article:

> There are discoverable uniformities in political behavior. These can be expressed in generalizations or theories with explanatory and predictive value. . . . The validity of such generalizations must be testable in principle, by reference to relevant behavior. . . . Precision in recording of data and the statement of findings requires measurement and quantification.

The rise of behavioralism reflected a frustration with the perceived subjectivity of traditional modes of sociological and political inquiry, particularly historical analysis and philosophical speculation. History and philosophy, behavioralists believed, led only to ideological bickering, not to unbiased knowledge or reliable solutions to problems. But behavioralism also had technological origins. It was spurred by the postwar arrival of digital computers, machines that promised to open new horizons in the collection and analysis of data on human behavior. Objectivity would replace subjectivity. Technology would replace ideology.

Today's neobehavioralism has also been inspired by advances in computer technology, particularly the establishment of vast data banks of information on people's behavior and the development of automated statistical techniques to parse the information. The MIT data scientist Alex Pentland, in his revealingly titled 2014 book *Social Physics*, offered something of a manifesto for the new behavioralism, using terms that, consciously or not, echoed what was heard in the early sixties:

> We need to move beyond merely describing social structure to building a causal theory of social structure. Progress in developing this represents steps toward what [neuroscientist] David Marr called a computational theory of behavior: a mathematical explanation of why society reacts as it does and how these reactions may (or may not) solve human problems. . . . Such a theory could tie together mechanisms of social interactions with our newly acquired massive amounts of behavior data in order to engineer better social systems.

As with their predecessors, today's neobehavioralists view the scientific analysis of large sets of data as a means of escaping subjective modes of sociological inquiry and the ideological baggage those modes often carry. "The importance of a science of social physics," Pentland suggested, goes beyond "its utility in providing accurate, useful mathematical predictions." It promises to provide "a language that is better than the old vocabulary of markets and classes, capital and production."

> Words such as "markets," "political classes," and "social movements" shape our thinking about the world. They are useful, of course, but they also represent overly simplistic thinking; they

therefore limit our ability to think clearly and effectively. [Big data offers] a new set of concepts with which I believe we can more accurately discuss our world and plan the future.

Zuckerberg will lose his bet, and Pentland and the other new behavioralists will not discover "a causal theory of social structure" that can be expressed in the pristine language of mathematics. Neobehavioralism will, like behavioralism before it, fall short of its lofty goals, even if it does provide new insights into social dynamics. Despite, or because of, their subjective messiness, history and philosophy will continue to play central roles in the exploration of what makes all of us tick. The end of ideology is not nigh.

But there is something that sets neobehavioralism apart from behavioralism. The collection of behavioral data today generates great commercial value along with its value in social research, and there's an inevitable tension between the data's scientific and commercial exploitation. That tension will shadow any attempt to, as Pentland put it, "engineer better social systems." Better for whom, and by what measure? Even if no fundamental mathematical law of social relationships is in the offing, the ability to monitor and manipulate those relationships will continue to provide rich profit potential. One suspects that Zuckerberg's dream of a unified theory of love may be inspired less by cupid than by cupidity.

<3S AND MINDS

September 11, 2015

IN THAT JUNE Q&A SESSION, Mark Zuckerberg also offered a peek into the future of interpersonal communication: "One day, I believe we'll be able to send full, rich thoughts to each other directly using technology. You'll just be able to think of something and your friends will immediately be able to experience it too if you'd like. This would be the ultimate communication technology."

Wow. That's really going to require some incredible impulse control. Your inner filter is going to have to kick in not between thought and expression, as it does now, but before the formation of the thought itself. I mean, would you really want to share your raw thought-stream with another person, even a friend? Zuck may want instantaneous thought-sharing, but I'm thinking there's going to have to be some kind of broadcast delay built into the system, like they have on talk radio. Otherwise, the interbrain highway is going to resemble something out of a Mad Max movie.

Then again, I suppose a delay mechanism would defeat the purpose of replacing speech with computerized brain-to-brain messaging. The point is to get rid of the hesitancy, the haltingness, that accompanies the formulation and expression of thoughts in words. As political scientist William Davies has pointed out, today's "purveyors of 'smart' technologies" seek to achieve the seamless machine-human communication that the cyberneticists of the Cold War era could only dream about. The goal is to create a system of "perfectly predictable interaction between individual and environment, in which nothing needs to be said along the way." Beyond the efficiency gains, Silicon Valley would stand to profit from such a system. By developing a proprietary brain-computer network that renders human cogitation fully machine-readable, the tech industry would be able to transmit, store, parse, and hence to own, the entirety of our thoughts. "Industrial capitalism privatized the means of production," Davies observed. "Digital capitalism seeks to privatize the means of communication."

That's already happening. In digitizing human expression, the protocols of social networks are beginning to alter speech to make it more amenable to machine transmission and interpretation. Think of Like buttons, or other forms of online communication that involve, say, tapping an icon or clicking a checkbox or selecting an option from a drop-down menu. They routinize, or formalize, the expression of judgment and affiliation and emotion, purging it of ambiguity and indirection, not to mention nuance. They shape speech to the needs of the computer network—and the computer network's owner.

"The speaking of language is part of an activity, or of a form of life," wrote Wittgenstein in *Philosophical Investigations*. If human language is bound up in living, if it is an expression of both sense and sensibility, then computers, being nonliving, having no sensibility, will have a very difficult time mastering "natural-language processing" beyond a certain rudimentary level. The best solution, if you have a need to get computers to "understand" human communication, may be to avoid the problem altogether. Instead of figuring out how to get computers to understand natural language, you get people to speak artificial language, the language of computers. A good way to start is with Like buttons and other standardized messaging protocols. A good next step would be to encourage people to express themselves not through messy assemblages of fuzzily defined words but through neat, formal symbols—emoticons or emoji, for instance. When we speak with emoji, we're speaking a language that machines can understand.

People like Mark Zuckerberg have always been uncomfortable with natural language. Now they can do something about it.

IN THE KINGDOM OF THE BORED, THE ONE-ARMED BANDIT IS KING

September 15, 2015

IT STILL FEELS A little shameful to admit to the fact, but what engages us more and more is not the content but the mechanism. The poet Kenneth Goldsmith, in a *Los Angeles Review of Books* essay, writes of a recent day when he felt an urge to listen to some music by the avant-garde American composer Morton Feldman:

> I dug into my MP3 drive, found my Feldman folder and opened it up. Amongst the various folders in the directory was one labeled "The Complete Works of Morton Feldman." I was surprised to see it there; I didn't remember downloading it. Curious, I looked at its date—2009—and realized that I must've grabbed it during the heyday of MP3 sharity blogs. I opened it to find 79 albums as zipped files. I unzipped three of them, listened to part of one, and closed the folder. I haven't opened it since.

The pleasure of listening to music was not as great as Goldsmith anticipated. He found more pleasure in manipulating music files. "Engaging with media in a traditional sense is often the last thing we do," he observes. "In the digital ecosystem, the apparatuses surrounding the artifact are more engaging than the artifact itself." It was once assumed that digitization would liberate cultural artifacts from their physical containers. We'd be able to enjoy the wine without the bottles. But what's actually happened is different. We've come, as Goldsmith says, "to prefer the bottles to the wine."

It's as though we find ourselves, suddenly, in a vast library, an infinite library, a library of Borgesian proportions, and we discover that what's of most interest to us is not the books on the shelves but the intricacies of the Dewey decimal system.

Goldsmith's experience reminded me of a passage in Simon Reynolds's book *Retromania*.* Reynolds describes what happened after he got his first iPod and started playing with the Shuffle function:

> Shuffle offered a reprieve from the problem of choice. Like everybody, at first I was captivated by it and, like everybody, had all those experiences with mysterious recurrences of artists and uncanny sequencings. The downside of shuffle soon revealed itself, though. I became fascinated with the mechanism itself, and soon was always wanting to know what was coming up next. It was irresistible to click onto the next random selection. . . . Soon I was listening to just the first fifteen seconds of every track; then, not listening at all.

"Really," Reynolds concluded, "the logical culmination would have been for me to remove the headphones and just look at the track display."

What is the great innovation of SoundCloud, the popular music-streaming service? It has little to do with music and everything to do with the visual enrichment of the track display. Who needs to listen to the song when one can watch the song unspool colorfully on the screen through all its sonic peaks and valleys, triggering the display of comments as it goes?

Whatever lies on the other side of the interface seems less and less consequential. The interface is the thing. The interface is the content.

Abundance breeds boredom. When there's no end of choices, each choice feels disappointing. Listening to or watching one thing means you're not listening to or watching all the other things you might be listening to or watching. Reynolds quotes a telling line from Karla

* See page 292 for a review of *Retromania*.

Starr's 2008 article "When Every Song Ever Recorded Fits on Your MP3 Player, Will You Listen to Any of Them?" Confessed Starr: "I find myself getting bored even in the middle of songs simply because I *can*."

And so, bored by the content, bored by the art, we become obsessed with the interface. We seek to master the mechanism's intricate, fascinating functions: downloading and uploading, archiving and cataloging, monitoring readouts and notifications, watching time counts, streaming and pausing and skipping, clicking buttons marked with hearts or uplifted thumbs, checking Like totals. We become culture's technicians. We become bureaucrats of experience.

Managing the complexities of the interface provides an illusion of agency while alleviating the agony of choice. In the end, as Reynolds puts it, fiddling with the mechanism "relieves you of the burden of desire itself"—a burden that grows ever more burdensome as options proliferate. And so you find that you're no longer a music fan; you're a jukebox aficionado.

As the manufacturers of digital slot machines have discovered, a well-designed interface induces obsession. It's not the winnings, or the losses, that keep the players feeding money into the slots; it's the joy of operating a highly responsive machine. In her book *Addiction by Design: Machine Gambling in Las Vegas*, Natasha Dow Schüll tells of meeting a video-poker player named Mollie in a casino:

> When I ask Mollie if she is hoping for a big win, she gives a short laugh and a dismissive wave of her hand. . . . "Today when I win—and I do win, from time to time—I just put it back in the machines. The thing people never understand is that *I'm not playing to win*."

Why, then, does she play? "To keep playing—to stay in that machine zone where nothing else matters."

What does it feel like to be in the machine zone? Mollie explains: "It's like being in the eye of a storm, is how I'd describe it. Your vision is clear on the machine in front of you but the whole world is spinning around you, and you can't really hear anything. You aren't really there—you're with the machine and that's all you're with."

In a world dense with stuff, a captivating interface is the perfect consumer good. It packages the very act of consumption as a product. Click by click, we consume our consuming.

The machine zone is where we spend much of our time these days. It extends well beyond the traditional diversions of media and entertainment and gaming. The machine zone surrounds us. You go for a walk, and you find that what inspires you is not the scenery or the fresh air or the physical pleasure of the exercise, but rather the mounting step count on your smartphone's exercise app. "If I go just a little farther," you tell yourself, glancing yet again at the screen, "the app will reward me with a badge."

The mechanism is more than beguiling. The mechanism knows you, and it cares about you. You give it your attention, and it tells you that your attention has not been wasted.

THESES
IN
TWEETFORM

FIRST SERIES (2012)

1. The complexity of the medium is inversely proportional to the eloquence of the message.

2. Hypertext is a more conservative medium than text.

3. The best medium for the nonlinear narrative is the linear page.

4. Twitter is a more ruminative medium than Facebook.

5. The introduction of digital tools has never improved an art form.

6. The returns on interactivity quickly turn negative.

7. In the material world, doing is knowing; in media, the opposite is the case.

8. Increasing the intelligence of a network tends to decrease the intelligence of those connected to it.

9. The one new art form spawned by the computer—the video game—is the computer's prisoner.

10. Personal correspondence grows less interesting as the speed of its delivery quickens.

11. Programmers are the unacknowledged legislators of the world.

12. One always regrets one's retweets.

13. The album cover turned out to be indispensable to popular music.

14. Abundance of information breeds delusions of knowledge among the unwary.

15. To thine own image be true.

16. No great work of literature could have been written in hypertext.

17. Social media is a palliative for underemployment.

18. The philistine appears ideally suited to the role of cultural impresario online.

19. Television became more interesting when people started paying for it.

20. Instagram shows us what a world without art looks like.

SECOND SERIES (2013)

21. Recommendation engines are the best cure for hubris.

22. Vines would be better if they were one second shorter.

23. Hell is other selfies.

24. Twitter has revealed that brevity and verbosity are not always antonyms.

25. Personalized ads provide a running critique of artificial intelligence.

26. Who you are is what you do between notifications.

27. Online is to offline as a swimming pool to a pond.

28. People in love leave the sparsest data trails.

29. YouTube fan videos are the living fossils of the original web.

30. Mark Zuckerberg: the Grigory Potemkin of our time.

THIRD SERIES (2014)

31. Every point on the internet is a center of the internet.

32. One's sense of solipsism intensifies as one's follower count grows.

33. A thing contains infinitely more information than its image.

34. A book has many pages; an ebook has but one.

35. If a hard drive is a soul, the cloud is the oversoul.

36. The essence of an event is the ghost in the recording.

37. A Snapchat message becomes legible only after it vanishes.

38. Authenticity is artifice perfected.

39. When we turn on a GPS system, we become cargo.

40. Google searches us.

FOURTH SERIES (2015)

41. Tools extend us; media confine us.

42. People take facts as metaphors; computers take metaphors as facts.

43. We need not fear robots until robots fear us.

44. Engineers are ethicists in denial.

45. Friction is virtuosity's lubricant.

46. A car without a steering wheel is comic; a car without a rearview mirror is tragic.

47. One feels lightest after one clears one's browser cache.

48. The things of the world manifest themselves as presence or absence.

49. Memory is the medium of absence; time is the medium of presence.

50. A bird resembles us most when it flies into a window.

THE EUNUCH'S CHILDREN

Essays and Reviews

FLAME AND FILAMENT

ONE OF THE GREATEST of all inventions was also one of the simplest: the wick. We don't know who first realized, many thousands of years ago, that fire could be isolated at the tip of a twisted piece of cloth and fed, through capillary action, by a reservoir of wax or oil, but the discovery was, as historian Wolfgang Schivelbusch writes in *Disenchanted Night*, "as revolutionary in the development of artificial lighting as the wheel in the history of transport." The wick tamed fire, allowing it to be used with a precision and an efficiency far beyond what was possible with a wooden torch or a bundle of twigs. In the process, it helped domesticate us as well. It's hard to imagine civilization progressing to where it is today by torchlight.

The wick proved an amazingly hardy creation. It remained the dominant lighting technology all the way to the nineteenth century, when it was replaced first by the wickless gas lamp and then, more decisively, by Edison's electricity-powered incandescent bulb with its glowing metal filament. Cleaner, safer, and even more efficient than the flame it replaced, the lightbulb was welcomed into homes and offices around the world. But along with its many practical benefits, electric light brought subtle and unexpected changes to the way people lived. The fireplace, the candle, and the oil lamp had been focal points of households. Fire was, as Schivelbusch puts it, "the soul of the house." Families would in the evening gather in a central room, drawn by the flickering flame, to chat about the day's events or otherwise pass time together. Electric light, together with central heat, dissolved that ancient tradition. Family members began to spend more time in different rooms in the evening, studying or reading or working alone. Each person gained more privacy, and a greater sense of autonomy, but the cohesion of the family weakened.

Cold and steady, electric light lacked the allure of the flame. It was not mesmerizing or soothing but strictly functional. It turned light into an industrial commodity. A German diarist in 1944, forced to use candles instead of lightbulbs during nightly air raids, was struck by the difference. "We have noticed," he wrote, "in the 'weaker' light of the candle, objects have a different, a much more marked profile—it gives them a quality of 'reality.'" This quality, he continued, "is lost in electric light: objects (seemingly) appear much more clearly, but in reality it *flattens* them. Electric light imparts too much brightness and thus things lose body, outline, substance—in short, their essence."

We're still attracted to a flame at the end of a wick. We light candles to set a romantic or a calming mood, to mark a special occasion. We buy ornamental lamps that are crafted to look like candleholders, with bulbs shaped as stylized flames. But we can no longer know what it was like when fire was the source of all light. The number of people who remember life before the arrival of Edison's bulb has dwindled to just a few, and when they go they'll take with them all remaining memory of that earlier, pre-electric world. The same will happen, sometime toward the end of this century, with the memory of the world that existed before the computer and the internet became commonplace. We'll be the ones who bear it away.

All technological change is generational change. The full power and consequence of a new technology are unleashed only when those who have grown up with it become adults and begin to push their outdated parents to the margins. As the older generations die, they take with them their knowledge of what was lost when the new technology arrived, and only the sense of what was gained remains. It's in this way that progress covers its tracks, perpetually refreshing the illusion that where we are is where we were meant to be.

From The Big Switch
2008

IS GOOGLE MAKING
US STUPID?

"Dave, stop. Stop, will you? Stop, Dave. Will you stop, Dave?"
So the supercomputer HAL pleads with the implacable astronaut Dave
Bowman in a famous and weirdly poignant scene toward the end of
Stanley Kubrick's *2001: A Space Odyssey*. Bowman, having nearly been
sent to a deep-space death by the malfunctioning machine, is calmly,
coldly disconnecting the memory circuits that control its artificial
brain. "Dave, my mind is going," HAL says, forlornly. "I can feel it. I
can feel it."

I can feel it, too. Over the past few years I've had an uncomfortable sense that someone, or something, has been tinkering with my
brain, remapping the neural circuitry, reprogramming the memory.
My mind isn't going—so far as I can tell—but it's changing. I'm not
thinking the way I used to think. I can feel it most strongly when I'm
reading. Immersing myself in a book or a lengthy article used to be
easy. My mind would get caught up in the narrative or the turns of
the argument, and I'd spend hours strolling through long stretches of
prose. That's rarely the case anymore. Now my concentration starts
to drift after two or three pages. I get fidgety, lose the thread, begin
looking for something else to do. I feel as if I'm always dragging my
wayward brain back to the text. The deep reading that used to come
naturally has become a struggle.

I think I know what's going on. For more than a decade now, I've
been spending a lot of time online, searching and surfing and sometimes adding to the great databases of the internet. The web has been
a godsend to me as a writer. Research that once required days in the
stacks or periodical rooms of libraries can now be done in minutes. A
few Google searches, some quick clicks on hyperlinks, and I've got the

telltale fact or pithy quote I was after. Even when I'm not working, I'm as likely as not to be foraging in the web's info-thickets—reading and writing emails, scanning headlines and blog posts, watching videos and listening to podcasts, or just tripping from link to link to link. (Unlike footnotes, to which they're sometimes likened, hyperlinks don't merely point to related works; they propel you toward them.)

For me, as for others, the net is becoming a universal medium, the conduit for most of the information that flows through my eyes and ears and into my mind. The advantages of having immediate access to such an incredibly rich store of information are many, and they've been widely described and duly applauded. "The perfect recall of silicon memory," *Wired*'s Clive Thompson has written, "can be an enormous boon to thinking." But that boon comes at a price. As the media theorist Marshall McLuhan pointed out in the 1960s, media are not just passive channels of information. They supply the stuff of thought, but they also shape the process of thought. And what the net seems to be doing is chipping away my capacity for concentration and contemplation. My mind now expects to take in information the way the net distributes it: in a swiftly moving stream of particles. Once I was a scuba diver in the sea of words. Now I zip along the surface like a guy on a jet ski.

I'm not the only one. When I mention my troubles with reading to friends and acquaintances—literary types, most of them—many say they're having similar experiences. The more they use the web, the more they have to fight to stay focused on long pieces of writing. Some of the bloggers I follow have also begun mentioning the phenomenon. Scott Karp, who writes a blog about online media, recently confessed that he has stopped reading books altogether. "I was a lit major in college, and used to be [a] voracious book reader," he wrote. "What happened?" He speculates on the answer: "What if I do all my reading on the web not so much because the way I read has changed, i.e. I'm just seeking convenience, but because the way I THINK has changed?"

Bruce Friedman, who blogs regularly about the use of computers in medicine, has also described how the internet has altered his mental habits. "I now have almost totally lost the ability to read and absorb a longish article on the web or in print," he wrote earlier this year. A pathologist who has long been on the faculty of the University of Michigan Medical School, Friedman elaborated on his comment in a telephone conversation with me. His thinking, he said, has taken on a "staccato" quality, reflecting the way he quickly scans short passages of text from many sources online. "I can't read *War and Peace* anymore," he admitted. "I've lost the ability to do that. Even a blog post of more than three or four paragraphs is too much to absorb. I skim it."

Anecdotes alone don't prove much. And we still await the long-term neurological and psychological experiments that will provide a definitive picture of how internet use affects cognition. But a recently published study of online research habits, conducted by scholars from University College London, suggests that we may well be in the midst of a sea change in the way we read and think. As part of the five-year research program, the scholars examined computer logs documenting the behavior of visitors to two popular research sites, one operated by the British Library and one by a U.K. educational consortium, that provide access to journal articles, ebooks, and other sources of written information. They found that people using the sites exhibited "a form of skimming activity," hopping from one source to another and rarely returning to any source they'd already visited. They typically read no more than one or two pages of an article or book before they would "bounce" to another site. Sometimes they'd save a long article, but there's no evidence that they ever went back and actually read it. The authors of the study report: "It is clear that users are not reading online in the traditional sense; indeed there are signs that new forms of 'reading' are emerging as users 'power browse' horizontally through titles, contents pages and abstracts going for quick wins. It almost seems that they go online to avoid reading in the traditional sense."

Thanks to the ubiquity of text on the internet, not to mention the popularity of text messaging on cell phones, we may well be reading more today than we did in the 1970s or 1980s, when television was our medium of choice. But it's a different kind of reading, and behind it lies a different kind of thinking—perhaps even a new sense of the self. "We are not only *what* we read," says Maryanne Wolf, a developmental psychologist at Tufts University and the author of *Proust and the Squid: The Story and Science of the Reading Brain.* "We are *how* we read." Wolf worries that the style of reading promoted by the net, a style that puts "efficiency" and "immediacy" above all else, may be weakening our capacity for the kind of deep reading that emerged when an earlier technology, the printing press, made long and complex works of prose commonplace. When we read online, she says, we tend to become "mere decoders of information." Our ability to interpret text, to make the rich mental connections that form when we read deeply and without distraction, remains largely disengaged.

Reading, explains Wolf, is not an instinctive skill for human beings. It's not etched into our genes the way speech is. We have to teach our minds how to translate the symbolic characters we see into the language we understand. And the media or other technologies we use in learning and practicing the craft of reading play an important part in shaping the neural circuits inside our brains. Experiments demonstrate that readers of ideograms, such as the Chinese, develop a mental circuitry for reading that is very different from the circuitry found in those of us whose written language employs an alphabet. The variations extend across many regions of the brain, including those that govern such essential cognitive functions as memory and the interpretation of visual and auditory stimuli. We can expect as well that the circuits woven by our use of the net will be different from those woven by our reading of books and other printed works.

SOMETIME IN 1882, Friedrich Nietzsche bought a typewriter—a Malling-Hansen Writing Ball, to be precise. His vision was failing, and keeping his eyes focused on a page had become exhausting and painful, often bringing on crushing headaches. He had been forced to curtail his writing, and he feared that he would soon have to give it up. The typewriter rescued him, at least for a time. Once he had mastered touch-typing, he was able to write with his eyes closed, using only the tips of his fingers. Words could once again flow from his mind to the page.

But the machine had a subtler effect on his work. One of Nietzsche's friends, a composer, noticed a change in the style of his writing. His already terse prose had become even tighter, more telegraphic. "Perhaps you will through this instrument even take to a new idiom," the friend wrote in a letter, noting that, in his own work, his " 'thoughts' in music and language often depend on the quality of pen and paper."

"You are right," Nietzsche replied, "our writing equipment takes part in the forming of our thoughts." Under the sway of the machine, writes the German media scholar Friedrich A. Kittler, Nietzsche's prose "changed from arguments to aphorisms, from thoughts to puns, from rhetoric to telegram style."

The human brain is malleable. People used to think that our mental meshwork, the dense connections formed among the hundred billion or so neurons inside our skulls, was largely fixed by the time we reached adulthood. But brain researchers have discovered that that's not the case. James Olds, a professor of neuroscience who directs the Krasnow Institute for Advanced Study at George Mason University, says that even the adult mind "is very plastic." Nerve cells routinely break old connections and form new ones. "The brain," according to Olds, "has the ability to reprogram itself on the fly, altering the way it functions."

As we use what the sociologist Daniel Bell has called our "intellectual technologies"—the tools that extend our mental rather than our physical capacities—we inevitably begin to take on the qualities of those technologies. The mechanical clock, which came into common use in the fourteenth century, provides a compelling example. In *Technics and Civilization*, the historian and cultural critic Lewis Mumford described how the clock "disassociated time from human events and helped create the belief in an independent world of mathematically measurable sequences." The "abstract framework of divided time" became "the point of reference for both action and thought."

The clock's methodical ticking helped bring into being the scientific mind and the scientific man. But it also took something away. As the late MIT computer scientist Joseph Weizenbaum observed in his 1976 book *Computer Power and Human Reason: From Judgment to Calculation*, the conception of the world that emerged from the widespread use of timekeeping instruments "remains an impoverished version of the older one, for it rests on a rejection of those direct experiences that formed the basis for, and indeed constituted, the old reality." In deciding when to eat, to work, to sleep, to rise, we stopped listening to our senses and started obeying the clock.

The process of adapting to new intellectual technologies is reflected in the changing metaphors we use to explain ourselves to ourselves. When the mechanical clock arrived, people began thinking of their brains as operating "like clockwork." Today, in the age of software, we have come to think of them as operating "like computers." But the changes, neuroscience tells us, go much deeper than metaphor. Thanks to our brain's plasticity, the adaptation occurs also at a biological level.

The internet promises to have particularly far-reaching effects on cognition. In a paper published in 1936, the British mathematician Alan Turing proved that a digital computer, which at the time existed only as a theoretical machine, could be programmed to perform the function of any other information-processing device. And that's what we're seeing today. The internet, an immeasurably powerful computing system, is subsuming most of our other intellectual technologies. It's becoming our map and our clock, our printing press and our typewriter, our calculator and our telephone, and our radio and TV.

When the net absorbs a medium, it re-creates that medium in its own image. It injects the medium's content with hyperlinks, blinking ads, and other digital gewgaws, and it surrounds the content with the content of all the other media it has absorbed. A new email message, for instance, may announce its arrival as we're glancing over the latest headlines at a newspaper's site. The result is to scatter our attention and diffuse our concentration.

The net's influence doesn't end at the edges of a computer screen, either. As people's minds become attuned to the crazy quilt of internet media, traditional media have to adapt to the audience's new expectations. Television programs add text crawls and pop-up ads, and magazines and newspapers shorten their articles, introduce capsule summaries, and crowd their pages with easy-to-browse info-snippets. When, in March of this year, the *New York Times* decided to devote the second and third pages of every edition to article abstracts, its design director, Tom Bodkin, explained that the "shortcuts" would give harried readers a quick "taste" of the day's news, sparing them the "less efficient" method of actually turning the pages and reading the articles. Old media have little choice but to play by the new-media rules.

Never has a communications system played so many roles in our lives—or exerted such broad influence over our thoughts—as the internet does today. Yet, for all that's been written about the net, there's been little consideration of how, exactly, it's reprogramming us. The net's intellectual ethic remains obscure.

ABOUT THE same time that Nietzsche started using his typewriter, an earnest young man named Frederick Winslow Taylor carried a stopwatch into the Midvale Steel plant in Philadelphia and began a historic series of experiments aimed at improving the efficiency of the plant's machinists. With the approval of Midvale's owners, he recruited a group of factory hands, set them to work on various metalworking machines, and recorded and timed their every movement as well as the operations of the machines. By breaking down every job into a sequence of small, discrete steps and then testing different ways of performing each one, Taylor created a set of precise instructions—an "algorithm," we might say today—for how each worker should work. Midvale's employees grumbled about the strict new regime, claiming that it turned them into little more than automatons, but the factory's productivity soared.

More than a hundred years after the invention of the steam engine, the industrial revolution had at last found its philosophy and its philosopher. Taylor's tight industrial choreography—his "system," as he liked to call it—was embraced by manufacturers throughout the country and, in time, around the world. Seeking maximum speed, maximum efficiency, and maximum output, factory owners used time-and-motion studies to organize their work and configure the jobs of their workers. The goal, as Taylor defined it in his celebrated 1911 treatise *The Principles of Scientific Management*, was to identify and adopt, for every job, the "one best method" of work and thereby to effect "the gradual substitution of science for rule of thumb throughout the mechanic arts." Once his system was applied to all acts of manual labor, Taylor assured his followers, it would bring about a restructuring not only of industry but of society, creating a utopia of perfect efficiency. "In the past the man has been first," he declared; "in the future the system must be first."

Taylor's system is still very much with us; it remains the ethic of industrial manufacturing. And now, thanks to the growing power that computer engineers and software coders wield over our intellectual lives, Taylor's ethic is beginning to govern the realm of the mind as well. The internet is a machine designed for the efficient and auto-mated collection, transmission, and manipulation of information, and its legions of programmers are intent on finding the "one best method"—the perfect algorithm—to carry out every mental move-ment of what we've come to describe as "knowledge work."

Google's headquarters, in Mountain View, California—the Google-plex—is the internet's high church, and the religion practiced inside its walls is Taylorism. Google, says its chief executive, Eric Schmidt, is "a company that's founded around the science of measurement," and it is striving to "systematize everything" it does. Drawing on the terabytes of behavioral data it collects through its search engine and other sites, it carries out thousands of experiments a day, according to the *Harvard Business Review*, and it uses the results to refine the algorithms that increasingly control how people find information and extract meaning from it. What Taylor did for the work of the hand, Google is doing for the work of the mind.

The company has declared that its mission is "to organize the world's information and make it universally accessible and useful." It seeks to develop "the perfect search engine," which it defines as something that "understands exactly what you mean and gives you back exactly what you want." In Google's view, information is a kind of commodity, a utilitarian resource that can be mined and processed with industrial efficiency. The more pieces of information we can "access" and the faster we can extract their gist, the more productive we become as thinkers.

Where does it end? Sergey Brin and Larry Page, the gifted young men who founded Google while pursuing doctoral degrees in computer science at Stanford, speak frequently of their desire to turn their search engine into an artificial intelligence, a HAL-like machine that might be connected directly to our brains. "The ultimate search engine is something as smart as people—or smarter," Page said in a speech a few years back. "For us, working on search is a way to work on artificial intelligence." In a 2004 interview with *Newsweek*, Brin said, "Certainly if you had all the world's information directly attached to your brain, or an artificial brain that was smarter than your brain, you'd be better off." Last year, Page told a convention of scientists that Google is "really trying to build artificial intelligence and to do it on a large scale."

Such an ambition is a natural one, even an admirable one, for a pair of math whizzes with vast quantities of cash at their disposal and a small army of computer scientists in their employ. A fundamentally scientific enterprise, Google is motivated by a desire to use technology, in Eric Schmidt's words, "to solve problems that have never been solved before," and artificial intelligence is the hardest problem out there. Why wouldn't Brin and Page want to be the ones to crack it?

Still, their easy assumption that we'd all "be better off" if our brains were supplemented, or even replaced, by an artificial intelligence is unsettling. It suggests a belief that intelligence is the output of a mechanical process, a series of discrete steps that can be isolated, measured, and optimized. In Google's world, the world we enter when we go online, there's little place for the fuzziness of contemplation. Ambiguity is not an opening for insight but a bug to be fixed. The human brain is just an outdated computer that needs a faster processor and a bigger hard drive.

The idea that our minds should operate as high-speed data-processing machines is not only built into the workings of the internet; it is the network's reigning business model. The faster we surf across the web—the more links we click and pages we view—the more opportunities Google and other companies gain to collect information about us and to feed us advertisements. Most of the proprietors of the commercial internet have a financial stake in collecting the crumbs of data we leave behind as we flit from link to link—the more crumbs, the better. The last thing these companies want is to encourage leisurely reading or slow, concentrated thought. It's in their economic interest to drive us to distraction.

MAYBE I'M just a worrywart. Just as there's a tendency to glorify technological progress, there's a countertendency to expect the worst of every new tool or machine. In Plato's *Phaedrus*, Socrates bemoaned the development of writing. He feared that, as people came to rely on the written word as a substitute for the knowledge they used to carry inside their heads, they would, in the words of one of the dialogue's characters, "cease to exercise their memory and become forgetful." And because they would be able to "receive a quantity of information without proper instruction," they would "be thought very knowledgeable when they are for the most part quite ignorant." They would be "filled with the conceit of wisdom instead of real wisdom." Socrates wasn't wrong—the new technology did often have the effects he feared—but he was shortsighted. He couldn't foresee the many ways that writing and reading would serve to spread information, spur fresh ideas, and expand human knowledge (if not wisdom).

The arrival of Gutenberg's printing press, in the fifteenth century, set off another round of teeth gnashing. The Italian humanist Hieronimo Squarciafico worried that the easy availability of books would lead to intellectual laziness, making men "less studious" and weakening their minds. Others argued that cheaply printed books and broadsheets would undermine religious authority, demean the work of scholars and scribes, and spread sedition and debauchery. As New York University professor Clay Shirky notes, "Most of the arguments made against the printing press were correct, even prescient." But, again, the doomsayers were unable to imagine the myriad blessings that the printed word would deliver.

So, yes, you should be skeptical of my skepticism. Perhaps those who dismiss critics of the internet as Luddites or nostalgists will be proved correct, and from our hyperactive, data-stoked minds will spring a golden age of intellectual discovery and universal wisdom. Then again, the net isn't the alphabet, and although it may replace the printing press, it produces something altogether different. The kind of deep reading that a sequence of printed pages promotes is valuable not just for the knowledge we acquire from the author's words but for the intellectual vibrations those words set off within our own minds. In the quiet spaces opened up by the sustained, undistracted reading of a book, or by any other act of contemplation, for that matter, we make our own associations, draw our own inferences and analogies, foster our own ideas. Deep reading, as Maryanne Wolf argues, is indistinguishable from deep thinking.

If we lose those quiet spaces, or fill them up with "content," we will sacrifice something important not only in our selves but in our culture. In a recent essay, the playwright Richard Foreman eloquently described what's at stake: "I come from a tradition of Western culture, in which the ideal (my ideal) was the complex, dense and 'cathedral-like' structure of the highly educated and articulate personality—a man or woman who carried inside themselves a personally constructed and unique version of the entire heritage of the West." But now, he continued, "I see within us all (myself included) the replacement of complex inner density with a new kind of self—evolving under the pressure of information overload and the technology of the 'instantly available.'"

As we are drained of our "inner repertory of dense cultural inheritance," Foreman concluded, we risk turning into " 'pancake people'—spread wide and thin as we connect with that vast network of information accessed by the mere touch of a button."

I'm haunted by that scene in *2001*. What makes it so poignant, and so weird, is the computer's emotional response to the disassembly of its mind: its despair as one circuit after another goes dark, its childlike pleading with the astronaut—"I can feel it. I can feel it. I'm afraid"—and its final reversion to what can only be called a state of innocence. HAL's outpouring of feeling contrasts with the emotionlessness that characterizes the human figures in the film, who go about their business with an almost robotic efficiency. Their thoughts and actions feel scripted, as if they're following the steps of an algorithm. In the world of *2001*, people have become so machinelike that the most human character turns out to be a machine. That's the essence of Kubrick's dark prophecy: As we come to rely on computers to mediate our understanding of the world, it is our own intelligence that flattens into artificial intelligence.

From The Atlantic
2008

SCREAMING FOR QUIET

IN 1906, JULIA BARNETT RICE, a wealthy New York physician and philanthropist, founded the Society for the Suppression of Unnecessary Noise. Rice, who lived with her husband and six children in a Manhattan mansion overlooking the Hudson River, had become enraged at the way tugboats would blow their horns incessantly while steaming up and down the busy waterway. During a typical night, the tugs would emit two or three thousand toots, most of which served merely as sonic greetings between friendly captains.

Armed with research documenting the health problems caused by the sleep-shattering racket, Rice launched a one-woman lobbying campaign that took her to police stations, health departments, the offices of shipping regulators, and ultimately the halls of Congress. Initially ignored, her pleas finally reached sympathetic ears in Washington—and she won her fight. New York and other East Coast cities placed tough new restrictions on the blowing of horns and whistles by tugs. Nights became much quieter, and a lot more restful.

Encouraged by the victory, Rice organized her quiet-promoting society and proceeded to attack, and hush, other producers of what we would today call noise pollution. One of the organization's most celebrated accomplishments was its successful campaign, heartily backed by Mark Twain, to get school kids to pledge to keep quiet when walking or playing near hospitals. In the long history of antinoise crusaders, Rice stands as one of the few that have actually made a difference. But even her string of successes soon ended. The Society for the Suppression of Unnecessary Noise fell by the wayside when a powerful new noisemaker—the automobile—made its way into cities. The roar of motorized traffic quickly drowned out the protests of do-gooders.

As George Prochnik reveals in *In Pursuit of Silence*, his genial study of the noisiness of modern life, the story of the Society for the Suppression of Unnecessary Noise comes with a wry, and revealing, punch line. It seems that the fellow who drove the first motorcar in Manhattan was none other than Julia Rice's husband, Isaac, who was particularly fond of racing his new runabout through the once tranquil byways of Central Park. The big challenge in fighting unnecessary noise was—and still is—that the people making the din rarely find it unnecessary. One person's cacophony is another's joyride.

Automobile traffic has today become the most pervasive and noxious source of noise in the world. Prochnik cites statistics from the World Health Organization indicating that the sounds emitted by cars' engines, brakes, and tires actually cause significantly more illness than the exhaust spewing from their tailpipes. The stress and the sleeplessness resulting from traffic noise take a particular toll on the heart, contributing to many thousands of fatal heart attacks every year. And yet, in another sign of the subjectivity of our experience of sound, when people are surveyed about the noises they find most disturbing, they point not to traffic but to the barky dog in a nearby yard or the raucous late-night party down the street. We tune out civilization's ever-present racket but find unendurable our neighbors' occasional pleasures and excesses.

We have adapted so well to the noisiness that surrounds us that we rarely even think of it as a problem. Julia Rice-style protests have become rare. Today, in fact, most urban noise-control programs have little to do with making places quieter. They are aimed instead at designing "soundscapes" that make a city's clamor a little more agreeable by, for example, tweaking the way traffic flows through streets or the way voices spill from bars and restaurants. Soundscaping has also become popular among retailers, who now eschew generic Muzak in favor of meticulously customized store soundtracks that reinforce their brand image while propelling shoppers toward cash registers. A sound designer tells Prochnik that the thunderous beats pumped out by the sound systems in Abercrombie & Fitch outlets are engineered to create "a state of celebratory arousal" that reaches its climax with the purchase of a hoodie.

Soundscaping occurs at a more intimate level as well. When we want to isolate ourselves from society's ambient noise, we rarely think to seek out quiet spots. Instead, we just crank up our own personal volume knob. To make sure that we can drown out traffic noise during commutes, we upgrade the sound systems in our cars to include powerful amplifiers and subwoofers. Fighting fire with fire, at home we turn up our televisions and stereos to mask street noise—and the barking of the neighbor's dog.

The most popular of contemporary sound-management tools is, by far, the ubiquitous iPod. As soon as we plug the cute white earbuds into our ear canals, we enter the refuge of a personally engineered soundscape. Aural experience becomes completely customized. The iPod doesn't just shield us from the sound of urban infrastructure. It also, as Prochnik writes, blocks out "the discretionary din that got plastered on top of that layer"—the din created by people talking on cell phones, playing video games, and listening to their own iPods. All of us are now combatants in a sonic arms race, with no end in sight.

Unfortunately, our bodies are ill suited to the loudness we wrap ourselves in. Human ears, like those of other animals, evolved in a world that put a premium on keeping quiet. Although a well-timed roar might now and then scare off a predator, survival more often hinged on the ability to hear the movements of others without being heard yourself. The ear, which in its original form was a vibration-sensing part of the jaw, developed into an incredibly sensitive amplifier, able, quite literally, to hear a pin drop. One expert on the biology of hearing tells Prochnik that a sound gets a hundred times louder between the moment it enters our ears and the moment we actually hear it. That physiological amplification is extremely useful when it comes to getting an advance warning on the approach of a predator during a quiet night, but it becomes a disability in our age of iPods, boom cars, and blaring TVs. Hearing loss, Prochnik reports, is epidemic, and yet, seeking refuge from noise in more noise, we continue to jack up the volume.

Prochnik is adept at reporting on the work of scientists, soldiers, and soundproofers—people for whom noise is a decidedly earthly concern. He becomes less sure-footed when he travels to more ethereal realms—monasteries, Zen gardens—and tries to explain the attractions of silence. "When we confront silence," he muses during a visit with Trappist monks, "the mind reaches outward." Five pages later, he remarks that "the pursuit of silence often turns us deeper and deeper inward." Both statements may be true, but it would have been nice to learn how silence can push us in opposite metaphysical directions simultaneously. As it is, we're left hanging between platitudes.

Then again, maybe this helps explain why society has been getting ever noisier. We have little problem making the case for the necessity of noise as a product or byproduct of useful and entertaining technologies and pastimes. But when it comes to describing the benefits of silence, words fail us.

From The New Republic
2010

THE DREAMS OF READERS

"Spermatic." There's a word you don't come across much anymore. Not only does it sound fusty and arcane, as if it had been extracted from the nether regions of a moldy physiology handbook, but it seems fatally tainted with political incorrectness. Only the rash or the drunken would dare launch the word into a conversation at a cocktail party.

It wasn't always a pariah. In an essay published in the *Atlantic Monthly* in 1858, Ralph Waldo Emerson chose the adjective to describe the experience of reading: "I find certain books vital and spermatic, not leaving the reader what he was." For Emerson, the best books— the "true ones"—"take rank in our life with parents and lovers and passionate experiences, so medicinal, so stringent, so revolutionary, so authoritative." Books are not only alive; they give life, or at least give it a new twist.

Emerson drew a distinction between his idea of reading and one expressed a few centuries earlier by the French essayist Michel de Montaigne, who termed books "a languid pleasure." What was medicine for Emerson was wine to Montaigne. If my own experience is any guide, both men had it right. Like Montaigne, I've spent many happy hours under the spell of books, intoxicated by the beauty or wit of the prose, the plot's intrigue, the elegance of the argument. But there have also been times when, like Emerson, I've felt a book's metamorphic force, when reading becomes a means not just of diversion or enlightenment but of regeneration. One closes such a book a different person from the one who opened it. In his poem "Two Tramps in Mud Time," Robert Frost, one of Emerson's many heirs, wrote of the rare moments in life when

> *love and need are one,*
> *And the work is play for mortal stakes.*

That seems to me a good description of reading at its most vital and spermatic.

My life has been punctuated by books. *The Lord of the Rings* and *The Martian Chronicles* added mystery to my boyhood, opening frontiers to wander in and marvel at far beyond my suburban surroundings. The tumult of my teenage years was fueled by rock records, but it was put into perspective by books as various as Kerouac's *On the Road* and Hemingway's *In Our Time*, Philip Roth's *Portnoy's Complaint* and Joseph Heller's *Something Happened*. During my twenties, a succession of thin volumes of verse—Ted Hughes's *Lupercal*, Philip Larkin's *The Whitsun Weddings*, Seamus Heaney's *North*—were the wedges I used to pry open new ways of seeing and feeling. The list goes on: Hardy's *The Return of the Native*, Joyce's *Ulysses*, Elizabeth Bishop's *Poems*, Cormac McCarthy's *Blood Meridian*, Joan Didion's *The White Album*, Denis Johnson's *Jesus' Son*. Who would I be without those books? Someone else.

PSYCHOLOGISTS AND neurobiologists have begun studying what goes on in our minds as we read literature, and what they're discovering lends scientific weight to Emerson's observation. One of the trailblazers in this field is Keith Oatley, a cognitive psychologist at the University of Toronto and the author of several novels, including the acclaimed *The Case of Emily V.* "For a long time," Oatley told the Canadian magazine *Quill & Quire*, "we've been talking about the benefits of reading with respect to vocabulary, literacy, and these such things. We're now beginning to see that there's a much broader impact." A work of literature, particularly narrative literature, takes hold of the brain in curious and powerful ways. In his 2011 book *Such Stuff as Dreams: The Psychology of Fiction*, Oatley explained that "we don't just respond to fiction (as might be implied by the idea of reader response), or receive it (as might be implied by reception studies), or appreciate it (as in art appreciation), or seek its correct interpretation (as seems sometimes to be suggested by the New Critics). We create our own version of the piece of fiction, our own dream, our own enactment." Making sense of what transpires in a book's imagined reality appears to depend on "making a version of the action ourselves, inwardly."

One intriguing study, conducted a few years ago by research psychologists at Washington University in St. Louis, illuminates Oatley's point. The scholars used brain scans to examine the cellular activity that occurs inside people's heads as they read stories. They found that "readers mentally simulate each new situation encountered in a narrative." The groups of nerve cells, or neurons, activated in readers' brains "closely mirror those involved when [people] perform, imagine, or observe similar real-world activities." When, for example, a character in a story puts a pencil down on a desk, the neurons that control muscle movements fire in a reader's brain. When a character goes through a door to enter a room, electrical charges begin to flow through the areas in a reader's brain that are involved in spatial representation and navigation.

More than mere replication is going on. The reader's brain is not just a mirror. The actions and sensations portrayed in a story, the researchers wrote, are woven together "with personal knowledge from [each reader's] past experiences." Every reader of a book creates, in Oatley's terms, his own dream of the work—and he inhabits that dream as if it were an actual place.

When we open a book, it seems that we really do enter, as far as our brains are concerned, a new world—one conjured not just out of the author's words but out of our own memories and desires—and it is our cognitive immersion in that world that gives reading its emotional force. Psychologists draw a distinction between two kinds of emotions that can be inspired by a work of art. There are the "aesthetic emotions" we feel when we view art from a distance, as a spectator: a sense of beauty or of wonder, for instance, or a feeling of awe at the artist's craft or the work's unity. These are the emotions that Montaigne likely had in mind when he spoke of reading's languid pleasure. And then there are the "narrative emotions" we experience when, through the sympathetic actions of our nervous system, we become part of a story, when the distance between the attendee and the attended evaporates. These are the emotions Emerson may have had in mind when he described the spermatic, life-giving force of a "true book."

Readers routinely speak of how books have changed them. A 1999 survey of people who read for pleasure found that nearly two-thirds believe they have been transformed in lasting ways by their reading. This is no fancy. Experiencing strong emotions has been shown to cause alterations in brain functions, and that appears to hold true for the feelings stirred by stories. "The emotions evoked by literary fiction," Oatley reported in a 2010 paper written with psychologist Raymond Mar of York University in Toronto, "have an influence on our cognitive processing after the reading experience has ended." Although the full extent of that influence has yet to be measured in a laboratory, and may never be, it seems likely that the unusual length of time that a typical reader spends immersed in the world of a book would result in particularly strong emotional responses and, in turn, cognitive changes. These effects would be further amplified, argued Oatley and Mar, by the remarkably "deep simulation of experience that accompanies our engagement with literary narratives."

A 2009 experiment conducted by Oatley and three colleagues suggests that the emotions stirred by literature can even alter, in subtle but real ways, people's personalities. The researchers recruited 166 university students and gave them a standard personality test measuring such traits as extraversion, conscientiousness, and agreeableness. One group of the subjects read the Chekhov story "The Lady with the Dog," while a control group read a synopsis of the story's events, stripped of its literary qualities. Both groups then took the personality test again. The results revealed that the people "who read the short story experienced significantly greater change in personality than the control group," and the effect appeared to be tied to the strong emotional response that the story provoked. What was really interesting, Oatley says, is that the readers "all changed in somewhat different ways." A book is rewritten in the mind of every reader, and the book rewrites each reader's mind in a unique way, too.

What is it about literary reading that gives it such sway over how we think and feel, and maybe even who we are? Norman Holland, a scholar at the McKnight Brain Institute at the University of Florida, has been studying literature's psychological effects for many years, and he offers a provocative answer to that question. Although our emotional and intellectual responses to events in literature mirror, at a neuronal level, the responses we would feel if we actually experienced those events, the mind we read with, Holland argues in his book *Literature and the Brain*, is a very different mind from the one we use to navigate the real world. In our day-to-day routines, we are always trying to manipulate or otherwise act on our surroundings, whether it's by turning a car's steering wheel or frying an egg or tapping a button on a smartphone. But when we open a book, our expectations and attitudes change. Because we understand that "we cannot or will not change the work of art by our actions," we are relieved of our desire to exert an influence over objects and people and hence can "disengage our [cognitive] systems for initiating actions." That frees us to become absorbed in the imaginary world of the literary work. We read the author's words with "poetic faith," to borrow a phrase that the psychologically astute Samuel Coleridge used two centuries ago.

"We gain a special trance-like state of mind in which we become unaware of our bodies and our environment," explains Holland. "We are 'transported.'" It is only when we leave behind the incessant busyness of our lives in society that we open ourselves to literature's regenerative power. That doesn't mean that reading is antisocial. The central subject of literature is society, and when we lose ourselves in a book we often receive an education in the subtleties and vagaries of human relations. Several studies have shown that reading tends to make us at least a little more empathetic, a little more alert to the inner lives of others. A series of experiments by researchers at the New School for Social Research, reported in *Science* in 2013, showed that reading literary fiction can strengthen a person's "theory of mind," which is what psychologists call the ability to understand what other people are thinking and feeling. "Fiction is not just a simulator of a social experience," one of the researchers, David Comer Kidd, told *The Guardian*; "it is a social experience." The reader withdraws in order to connect more deeply.

THE DISCOVERIES about literature's psychological and cognitive effects won't come as a surprise to readers. The research will serve mainly to confirm their intuitions. But the science is important nonetheless. It arrives at a crucial moment in the history of reading, and perhaps of literature. Not only has a new medium for reading—the computer screen—become popular as an alternative to the printed page; the value of reading as an end in itself is coming in for questioning. A strangely distorted view of reading has gained currency in some quarters. A group of social networking enthusiasts has taken to referring to the book, in its traditional form, as a "passive" medium, lacking the "interactivity" of websites, apps, and video games. Because a page of paper can't accommodate links, Like buttons, search boxes, comment forms, and all the other spurs to online activity we've become accustomed to, the reasoning goes, the readers of books must be mere consumers of content, inert caricatures of Montaigne's languid reader. Jeff Jarvis, a media consultant who teaches journalism at the City University of New York, gave voice to this way of thinking in a post on his blog. Claiming that printed pages "create, at best, a one-way relationship with a reader," he concluded that, in the internet era, "the book is an outdated means of communicating information." He declared that "print is where words go to die."

Anyone who would reduce a book to a "means of communicating information," as if it were a canister for shuttling facts and figures among bureaucrats, is probably not the best guide to the possibilities of literary experience. But when foolish ideas get into progress's slipstream, they can travel far. Some makers of e-reading devices and related software applications are embracing the notion that literature could do with a digital upgrade, that the experience of reading would improve if it were less solitary and more "social." Books "often live a vibrant life offline," grants a Google executive, but they will be able to "live an even more exciting life online." Such views reflect more than just technological enthusiasm. Something deeper is going on. Society is growing ever more skeptical of the value of solitude, ever more suspicious of even the briefest of withdrawals into inactivity and apparent purposelessness. We see it in the redefinition of receptive states of mind as passive states of mind. We see it in an education system that seems uncomfortable with any "outcome" unsuited to formal measurement. We see it in the self-contempt of the humanities. We see it in the glorification of the collaborative team and the devaluation of the self-reliant individual. We see it in the general desire to make all experience interactive and transactional.

In a 2003 lecture, Andrew Louth, a theology professor at the University of Durham in England, drew a distinction between "the free arts" and "the servile arts." The servile arts, he said, are those "to which a man is bound if he has in mind a limited task." They are the arts of production and consumption, of getting stuff done, to which most of us devote most of our waking hours. The free arts, among which Louth included reading as well as contemplation and prayer, are those characterized, in one way or another, by "the search for knowledge for its own sake." They are aimed at no useful or measurable end, and by engaging in them we slip, if only briefly, the bonds of the practical. We open ourselves to aesthetic and spiritual possibilities. We embrace and inhabit an ideal that was once central to the idea of culture itself: "that there is more to human life than a productive, well-run society." This ideal, Louth said, is now "under serious threat, and with it our notion of civilization."

The computer exists to aid in transactions. It is never not processing. As society becomes more narrowly focused on the servile arts, the computer naturally becomes more central to its operation. The relationship becomes symbiotic, and the free arts, which are antithetical to the transactional, are pushed to the margin. Then again, it might be argued that the margin has always been the best place to relax with a book. As Norman Holland suggests, the deepest kinds of reading entail a dampening of our urge to act. They require a withdrawal from quotidian busyness and so are marginal to modern society almost by definition. Montaigne's and Emerson's views may actually be more in concert than in conflict. It may be that readers have to enter a state of languid pleasure, a dream, before they can experience the full spermatic vitality of a book. Far from being a sign of passivity, the reader's outward repose signals the most profound kind of inner activity, the kind that goes unregistered by society's sensors.

From Stop What You're Doing and Read This!
2012

LIFE, LIBERTY, AND THE PURSUIT OF PRIVACY

IN A 1963 SUPREME COURT opinion, Chief Justice Earl Warren warned that "the fantastic advances in the field of electronic communication constitute a great danger to the privacy of the individual." The advances have continued in the decades since, and the danger has grown. Today, as companies strive to personalize the services and advertisements they provide over the internet, the surreptitious collection of personal information is rampant. The very idea of privacy is under threat.

Most of us view personalization and privacy as desirable things, and we understand that enjoying more of one means giving up some of the other. To have goods, services, and promotions tailored to our circumstances and desires, we need to divulge information about ourselves to corporations, governments, or other outsiders. Such tradeoffs have always been part of our lives as consumers and citizens. But now, thanks to the net, we're losing our ability to understand and control these tradeoffs—to choose, consciously and with awareness of the consequences, what information about ourselves we disclose and what we don't. Incredibly detailed data about our lives are being harvested from online databases without our knowledge, much less our approval.

Even though the internet is a very social place, we often access it in seclusion. We assume that we're anonymous as we go about our business online. As a result, we treat the net not just as a shopping mall and a library but as a personal diary and, sometimes, a confessional. Through the sites we visit and the searches we make, we disclose details not only about our jobs, hobbies, families, politics, and health, but also about our secrets, fantasies, even our peccadilloes. The sense of anonymity is an illusion. Pretty much everything we do online, down to individual keystrokes and clicks, is recorded, stored in cookies and corporate databases, and connected to our identities, either explicitly through our usernames, credit card numbers, and the IP addresses assigned to our computers, or implicitly through our searching, surfing, and purchasing histories.

A few years ago, a computer consultant named Tom Owad published the results of an experiment that provided a chilling lesson in just how easy it is to extract sensitive personal data from the net. Owad wrote a simple piece of software that allowed him to download the wish lists that Amazon.com customers use to catalog products that they plan to purchase or would like to receive as gifts. These lists usually include the customer's name and his or her city and state. Using a couple of standard-issue PCs, Owad was able to copy more than a quarter million wish lists over the course of a day. He searched his new database for controversial or politically sensitive books and authors, from Vonnegut's *Slaughterhouse-Five* to the Koran. He then used Yahoo People Search to identify addresses and phone numbers for many of the list keepers.

Owad ended up with maps of the United States showing the locations of people interested in particular books and ideas, including George Orwell's *1984*. He could just as easily have published a map showing the residences of people interested in books about treating depression or adopting a child or growing marijuana. "It used to be," he concluded, "you had to get a warrant to monitor a person or a group of people. Today, it is increasingly easy to monitor ideas. And then track them back to people."

What Owad did by hand can be performed automatically, with data-mining software that draws from many sites and databases. One of the essential characteristics of the net is the interconnection of diverse stores of information. The "openness" of databases is what gives the system much of its usefulness. But it also makes it easy to discover hidden relationships among far-flung bits of data.

In 2006, a team of scholars from the University of Minnesota described how easy it is for data-mining software to generate detailed profiles of individuals—even when they post information anonymously. The software is based on a simple principle: People tend to leave lots of little pieces of information about themselves and their opinions in many different places on the web. By identifying correspondences among the data, sophisticated algorithms can identify individuals with extraordinary precision. And it's not a big leap from there to discovering the people's names. The researchers noted that most Americans can be identified by name and address using only their ZIP code, birthday, and gender—three pieces of information that people often divulge when they register at a website.

The more deeply the net gets woven into our work and leisure, the more exposed we become. Over the last few years, as social networking services have grown in popularity, people have come to entrust ever more intimate details about their lives to sites like Facebook and Twitter. The incorporation of GPS transmitters into cell phones and the rise of location-tracking services like Foursquare provide powerful tools for assembling moment-by-moment records of people's movements. As reading shifts from printed pages onto networked devices like the Kindle and the Nook, it becomes possible for companies to more closely monitor people's reading habits—down to the amount of time they spend on a page.

"You have zero privacy," Scott McNealy remarked back in 1999, when he was chief executive of Sun Microsystems. "Get over it." Other Silicon Valley CEOs have expressed similar sentiments in just the last few months. "If you have something that you don't want anyone to know, maybe you shouldn't be doing it in the first place," Google's Eric Schmidt said in December, in response to a question about his company's collection of personal information. While internet companies may be complacent about the erosion of privacy—they, after all, profit from the trend—the rest of us should, as Justice Warren suggested, be wary.

An obvious and very real danger is that personal data could fall into the wrong hands. Data-mining tools are available not only to legitimate corporations and researchers, but also to crooks, con men, and creeps. As more data about us are collected and shared online, the threats from unsanctioned interceptions of the data grow. Criminal syndicates can use purloined information about our identities to commit financial fraud, and stalkers can use locational data to track our whereabouts. The first line of defense is simple common sense—to be cognizant and careful about what we disclose. But no amount of caution will protect us from the dispersal of information collected without our knowledge. If we're not aware of what data about us are available online, and how those data are being used, exchanged, and sold, it can be difficult to guard against abuses.

A more subtle risk is the possibility that personal information may be used to influence our behavior and even our thoughts in ways that are invisible to us. Personalization's evil twin is manipulation. As mathematicians and marketers refine data-mining algorithms, they gain more precise ways to predict people's behavior as well as how they'll react when they're presented with online ads and other digital stimuli.

As marketing pitches and product offerings become more tightly tied to our past patterns of behavior, they become more powerful as triggers of future behavior. Already, advertisers are able to infer extremely personal details about people by monitoring their web-browsing habits. They can then use that knowledge to create ad campaigns customized to particular individuals. A man who visits a site about obesity, for instance, may soon see a lot of promotional messages related to weight-loss treatments. A woman who does research about anxiety may be bombarded with pharmaceutical ads. The line between personalization and manipulation is a fuzzy one, but one thing is certain: We can never know if the line has been crossed if we're unaware of what companies know about us.

The greatest danger posed by the erosion of personal privacy is that it may lead us, as individuals and as a society, to devalue the concept of privacy, to see it as outdated and unimportant. We may begin to view privacy as something that gets in the way—a barrier to efficient shopping and socializing. That would be a tragedy. As the computer security expert Bruce Schneier has observed, privacy is not just a screen we hide behind when we do something naughty or embarrassing; privacy is intrinsic to liberty. When we feel that we're always being watched, we begin to lose our sense of self-reliance and free will and, along with it, our individuality. "We become children," writes Schneier, "fettered under watchful eyes."

Privacy is essential to life and liberty, and it's essential to the pursuit of happiness, in the broadest and deepest sense. We human beings are social creatures, but we're private creatures, too. What we don't share is as important as what we do share. The way the boundary between public self and private self is defined will vary greatly from person to person, which is exactly why it's so important to be vigilant in defending everyone's right to set that boundary as he or she sees fit.

From the Wall Street Journal
2010

HOOKED

TOM BISSELL IS A Renaissance Man for our out-of-joint times. He's not only a versatile and exuberant writer, a restless if ennui-ridden globetrotter, and a dedicated chewer of tobacco and smoker of pot, but he is, as well, a prodigiously gifted slayer of zombies and other digitized demons. A few pages into *Extra Lives*, his chronicle of a decade spent in the thrall of blood-and-guts video games, Bissell describes a fight he waged, in the guise of a perky, beret-wearing avatar named Jill, against a particularly resilient breed of undead in the early PlayStation classic Resident Evil. The scene opens with Jill wandering through the dining room of a spookily quiet mansion. After inspecting an ominous puddle of crimson fluid, she enters a hallway and finds herself cheek-to-jowl with a zombie. A "chewy struggle" ensues, which Jill survives only through some expert and gruesome knife work.

Released in 1996, Resident Evil inaugurated, as Bissell writes, a new era of "unbelievably brutal" games: "It provided gamers with one of the video-game form's first laboratories of virtual sadism, and I would be lying if I did not admit that it was, in its way, exhilarating." Both the sadism and the exhilaration would mount as video game consoles were outfitted with ever zippier computer chips and ever richer network connections. When we next find Bissell battling zombies, in Left 4 Dead, an Xbox 360 game released late in 2008, he is part of a four-man online team scrambling through a beautifully rendered "gauntlet of the damned." Blinded by a bile-spewing monster, Bissell abandons his grievously wounded teammates and takes refuge inside a safe house, locking the door behind him. But as he watches his buddies' health bars shrink away, he has a change of heart. He leaves the safe house, uses a shotgun to blast a slew of zombies back to Kingdom Come, and mounts a daring rescue of his comrades. The "heroic action" leaves a profound impression on him: "All the emotions I felt during those few moments—fear, doubt, resolve, and finally courage—were as intensely vivid as any I have felt while reading a novel or watching a film or listening to a piece of music."

Extra Lives is at its best in moments like these, when Bissell is actually playing the games he reports on. His descriptions of simulated gore and mayhem manage to be clinical, gripping, and hilarious all at once. He transmits to the reader the primitive, visceral excitements that make video games so enticing, even addictive, to their legions of devotees. One can almost understand why an intelligent, cultured man such as Bissell has been driven to dedicate large chunks of his adult life to bouts of gaming.

Much of this brief book, however, is dedicated to accounts not of playing video games but of discussing them. Bissell visits the offices of some of the top game makers—Epic, BioWare, Ubisoft—to talk about the mechanics of cut scenes, voice acting, and artificial intelligence systems. He chats about aesthetics with the iconoclastic designer Jonathan Blow, creator of the genre-bending "art game" Braid. He devotes one of his longest chapters to a report on the proceedings of an industry convention in Las Vegas in 2009, where "matters of narrative, writing, and story were discussed as though by a robot with a PhD in art semiotics from Brown."

Though they often have the feel of homework assignments written by a clever, slightly bored student, the journalistic sections of Bissell's book are illuminating and at times fascinating. They allow him to explore the challenges that game designers face as they struggle to expand the boundaries of their craft. Video games have become much more sophisticated in recent years—the spectacles they present are often tinged with moral ambiguity—but they continue to be plagued by what Bissell, in describing the Resident Evil series, calls "phenomenal stupidity." Their very form frustrates the ambitions of their creators.

The compromise in agency that lies at the heart of all games—control over the experience shifts clumsily between maker and player—

makes it difficult, perhaps impossible, for games to achieve the subtle and surprising emotional resonance that characterizes the finest works of art. Because games by definition have to be played, they can never be experienced with the combination of immersion and detachment, the repose, that characterizes the reader of a novel, the viewer of a painting, or the listener to a song or symphony.

Whatever their artistic talents and pretensions, game designers may in the end be fated to be toolmakers, creators of marvelous contraptions of intense but only passing interest. As Bissell's account makes clear, even the very best modern games—those with exquisite animation, smart writing, intriguing characters, and fresh story lines—have not been able to transcend their gameyness.

In his concluding chapter, Bissell turns his attention to one of the most loved and loathed of modern video game franchises: the sprawling, nihilistic underworld adventure Grand Theft Auto. It is also here, in the waning pages of the book, that the author makes a personal confession which ends up casting a weird shadow back over his entire enterprise. It turns out that, at the height of his passion for video games, when he was often playing Grand Theft Auto IV around the clock, Bissell was also snorting septum-dissolving quantities of cocaine.

At one point, he finds himself handing wads of cash to a Russian dealer in an alleyway in Tallinn, Estonia. "Soon," he recalls, "I was sleeping in my clothes. Soon my hair was stiff and fragrantly unclean." It's as if the demimonde of Grand Theft Auto has begun to leak into Bissell's own existence. Or maybe it's the other way around. Whose demons, the reader begins to wonder, has this gamer been slaying? The question, alas, goes largely unanswered. "Video games and cocaine," Bissell acknowledges, "feed on my impulsiveness, reinforce my love of solitude, and make me feel good and bad in equal measure." But that's as far as he goes in plumbing the meaning of his twin vices. The reader is left holding an enigma.

In the closing pages of *Extra Lives*, we see, if only faintly, the glimmerings of a deeper book, one that might have dissected the strange species of hopped-up man-boy whose compulsions find both spur and outlet in the hyperactive cartoon worlds rendered in meticulous detail on high-definition screens. But while Bissell may be fearless in fighting zombies, when it comes to exploring the murky and still unmapped territory of the adult gamer's soul, he loses his nerve. He stays in the safe house.

From The New Republic
2010

MOTHER GOOGLE

I TYPE THE LETTER "p" into Google's search box, and a list of ten suggested keywords, starting with "pandora" and concluding with "people magazine," appears beneath my cursor. I type an "r" after the "p," and the list refreshes itself. Now it begins with "priceline" and ends with "pregnancy calculator." I add an "o." The list updates again, going from "prom dresses" to "proxy sites."

Google is reading my mind—or trying to. Drawing on the terabytes of data it collects on people's search queries, it predicts, with each letter I type, what I'm most likely to be looking for. The company introduced the automatic recommendation of search terms in 2008, after a few years of testing. It's been tweaking the service, which it calls Google Suggest, ever since. This past spring, it rolled out the latest enhancement, which tailors suggestions to the searcher's whereabouts.

Google Suggest, and the similar services offered by other search engines, streamlines the discovery of information. When you click on a suggestion, you arrive at a page of search results, and the accompanying advertisements, a little faster than you would have had you typed out the query yourself. At a technical level, Google Suggest is remarkable. It testifies to the power of cloud computing—the serving up of software and information from big, distant data centers rather than from your computer's own hard drive. When I typed that first "p," the letter was beamed across the internet to a Google server in a building hundreds of miles away. The server read the letter, gathered ten popular search terms beginning with "p," and shot the list back to my screen. The intricate data processing exercise took less than a second. It felt magical.

It felt a little creepy, too. Every time Google presents me with a set of search terms customized to what I'm typing, it reminds me that I'm being watched. The company monitors my every keystroke. The privacy risks inherent in such long-distance exchanges of personal information became apparent in February when three European researchers revealed that they had used intercepts of some Google Suggest traffic to reconstruct people's searches. Alerted to the breach, Google quickly added a new layer of security to the transmissions, though the researchers claim that vulnerabilities remain.

I like Google—it's a cuddly company, and endlessly helpful—but I also resent it. It's like a nosy mother, intent on knowing everything her children are doing, and thinking. Worse, it's like a meddlesome mother, the helicoptering kind who can't let her kids do anything on their own. Start typing a keyword, and she immediately butts in, trying to finish it for you. At first you enjoy the solicitousness. But then you begin to bridle. You feel you're being smothered.

Matthew Crawford, in his book *Shop Class as Soulcraft*, writes eloquently about our modern affliction of "displaced agency." As corporations rush to anticipate our every need and preference, serving up a small selection of market-tested options the instant we require them, we are left with little room to act for ourselves. Our role narrows to making a selection from a set of choices that were chosen for us, based on some measure of popularity. "We have too few occasions to *do* anything," writes Crawford, "because of a certain predetermination of things from afar." Everything becomes easier, but less satisfying.

Software programmers are taking the displacement of personal agency to a new level. Intent on making their programs user-friendly, they're scripting the intimate processes of intellectual inquiry and even social attachment. We follow their scripts when we click on one of Google's keyword suggestions, and we follow them when we select from a list of categories to describe ourselves and our relationships on Facebook. These choices are convenient, but they're not our own.

They're generalizations masquerading as personalizations. To automate such decisions is to subcontract the construction of the self, or at least some part of it, to a business.

From The Atlantic
2010

THE LIBRARY OF UTOPIA

IN HIS 1938 BOOK *World Brain*, H. G. Wells imagined a time—not very distant, he sensed—when every person on the planet would have immediate access to "all that is thought or known." The thirties were a decade of rapid advances in microphotography, and Wells assumed that microfilm would be the technology to make the corpus of human knowledge universally available. "The time is close at hand," he wrote, "when any student, in any part of the world, will be able to sit with his projector in his own study at his or her convenience to examine *any* book, *any* document, in an exact replica."

Wells's optimism was misplaced. The Second World War put idealistic ventures on hold, and after peace was restored, technical constraints made his plan unworkable. Though microfilm would remain an important medium for storing and preserving documents, it proved too unwieldy, too fragile, and too expensive to serve as the basis for a broad system of knowledge transmission. But Wells's idea didn't die. Today, seventy-five years later, the prospect of creating a public repository of every book ever published—what the Princeton philosopher Peter Singer calls "the library of utopia"—seems well within our grasp. With the internet, we have an information system that can store and transmit documents efficiently and cheaply, delivering them on demand to anyone with a computer or a smartphone. All that remains to be done is to digitize the hundred million books that have appeared since Gutenberg invented movable type, index their contents, add some descriptive metadata, and put them online with tools for viewing and searching.

It sounds straightforward. And if it were just a matter of moving bits and bytes around, a universal online library might already exist. Google, after all, has been working on the challenge for ten years. But the search giant's book program has foundered; it is mired in a legal swamp. Now another momentous project to build a universal library is taking shape. It springs not from Silicon Valley but from Harvard University. The Digital Public Library of America—the DPLA—has big goals, big names, and big contributors. And yet for all the project's strengths, its success is far from assured. Like Google before it, the DPLA is learning that the major problem with constructing a universal library nowadays has little to do with technology. It's the thorny tangle of legal, commercial, and political issues that surrounds the publishing business. Internet or not, the world may still not be ready for the library of utopia.

LARRY PAGE isn't known for his literary sensibility, but he does like to think big. In 2002, the Google cofounder decided that it was time for his young company to scan all the world's books into its database. If printed texts weren't brought online, he feared, Google would never fulfill its mission of making the world's information "universally accessible and useful." After doing some book-scanning tests in his office—he manned the camera while Marissa Mayer, then a product manager, turned pages to the beat of a metronome—he concluded that Google had the smarts and the money to get the job done. He set a team of engineers and programmers to work. In a matter of months, they had invented an ingenious scanning device that used a stereoscopic infrared camera to correct for the bowing of pages that occurs when a book is opened. The new scanner made it possible to digitize books rapidly without cutting off their spines or otherwise damaging them. The team also wrote character recognition software that could decipher unusual fonts and other textual oddities in more than four hundred languages.

In 2004, Page and his colleagues went public with their project, which they would later name Google Book Search—a reminder that the company, at least originally, thought of the service essentially as an extension of its search engine. Five of the world's largest research libraries, including the New York Public Library and the libraries of Oxford and Harvard, signed on as partners. They agreed to let Google digitize books from their collections in return for copies of the images. The company went on a scanning binge, making digital replicas of millions of volumes. It didn't always restrict itself to books in the public domain; it scanned ones still under copyright, too. That's when the trouble started. The Authors Guild and the Association of American Publishers sued Google, claiming that copying entire books, even with the intent of showing only a few lines of text in search results, constituted "massive" copyright infringement.

Google then made a fateful choice. Instead of going to trial and defending Book Search on grounds that it amounted to "fair use" of copyright-protected material—a case that some legal scholars believe it might have won—it negotiated a sweeping settlement with its adversaries. In 2008, the company agreed to pay large sums to authors and publishers in return for permission to develop a commercial database of books. Under the terms of the deal, Google would be able to sell subscriptions to the database to libraries and other institutions while also using the service as a means for selling ebooks and displaying advertisements.

That only deepened the controversy. Librarians and academics lined up to oppose the deal. Many authors asked that their works be exempted from it. The U.S. Justice Department raised antitrust concerns. Foreign publishers howled. Last year, after a final round of legal maneuvering, federal district judge Denny Chin rejected the settlement, saying it "would simply go too far." Listing a variety of objections, he argued that the pact would not only "grant Google significant rights to exploit entire books, without permission of the copyright owners," but also reward the company for its "wholesale copying of copyrighted works" in the past. The company now finds itself nearly back at square one, with the original lawsuits slated to go to trial this summer. Facing new competitive threats from Facebook and other social networks, Google may no longer see Book Search as a priority. A decade after it began, Page's bold project has stalled.

———

IF YOU were looking for Larry Page's opposite, you would be hard pressed to find a better candidate than Robert Darnton. A distinguished historian and prize-winning author, a Chevalier in France's Légion d'Honneur, a 2011 recipient of the National Humanities Medal, the 72-year-old Darnton is everything that Page is not: patrician, diplomatic, and deeply embedded in the literary establishment. If Page is a bull in a china shop, Darnton is the china shop's proprietor.

But Darnton has one thing in common with Page: an ardent desire to see a universal library established online, a library that would, as he puts it, "make all knowledge available to all citizens." In the 1990s he initiated two groundbreaking projects to digitize scholarly and historical works, and by the end of the decade he was writing erudite essays about the possibilities of electronic books and digital scholarship. In 2007, Darnton was recruited to Harvard and named the director of its library system, giving him a prominent perch for promoting his dream. Although Harvard was one of the original partners in Google's scanning scheme, Darnton soon became the most eminent and influential critic of the Book Search settlement, writing articles and giving lectures in opposition to the deal. His criticism was as withering as it was learned. Google Book Search, he maintained, was "a commercial speculation" that, under the liberal terms of the settlement, seemed fated to grow into "a hegemonic, financially unbeatable, technologically unassailable, and legally invulnerable enterprise that can crush all competition." It would become "a monopoly of a new kind, not of railroads or steel, but of access to information."

Darnton's rhetoric seemed overwrought to some. University of Michigan librarian Paul Courant accused him of spreading "a dystopian fantasy." But Darnton had cause to be concerned. Over the years, he had watched commercial publishers ratchet up subscription prices for scholarly journals. Annual renewal fees had soared into the thousands of dollars for many periodicals, squeezing the budgets of research libraries. Darnton feared that Google, operating under the broad commercial protections granted by the settlement, would have the power to charge whatever it wanted for subscriptions to its database. Libraries might end up paying exorbitant sums to gain access to the very volumes they had let Google scan for free. The company's executives, Darnton acknowledged, seemed to be filled with idealism and goodwill, but there was no guarantee that they, or their successors, would not become profit-hungry predators in the future. By allowing "the commercialization of the content of our libraries," he argued, the agreement "would turn the Internet into an instrument for privatizing knowledge that belongs in the public sphere."

If libraries and universities worked together, Darnton argued, with funding from charitable foundations, they could build a true digital public library of America. Darnton's inspiration for the DPLA came not from today's technologists but from the great philosophers of the Enlightenment. As ideas circulated through Europe and across the Atlantic during the eighteenth century, propelled by the technologies of the printing press and the post office, thinkers like Voltaire, Rousseau, and Thomas Jefferson came to see themselves as citizens of a Republic of Letters, a freethinking meritocracy that transcended national borders. It was a time of great intellectual fervor and ferment, but the Republic of Letters was "democratic only in principle," Darnton pointed out in an essay in the *New York Review of Books*: "In practice, it was dominated by the wellborn and the rich."

With the internet, we could at long last rectify that inequity. By putting digital copies of works online, Darnton has argued, the collections of the country's great libraries could be made available to anyone with a computer and a link to the network. We could create a "Digital Republic of Letters" that would be truly free and open and democratic. The DPLA would allow us to "realize the Enlightenment ideals on which our country was founded."

HARVARD'S BERKMAN CENTER for Internet and Society accepted Darnton's challenge. It announced late in 2010 that it would lead an effort to build the DPLA and turn the Enlightenment dream into an Information Age reality. The project garnered seed money from the Alfred P. Sloan Foundation and attracted a steering committee that included a host of luminaries, including both Darnton and Courant as well as the chief librarian of Stanford University, Michael Keller, and the founder of the Internet Archive, Brewster Kahle. Named to chair the committee was John Palfrey, a young Harvard law professor who had written influential books about the internet.

The Berkman Center set an ambitious goal of having the digital library begin operating, at least in some rudimentary form, by April of 2013. Over the past year and a half, the project has moved quickly on several fronts. It has held public meetings to promote the library, solicit ideas, and recruit volunteers. It has organized six working groups to wrestle with various challenges, from defining its audience to resolving technical issues. And it has conducted an open competition, dubbed a "beta sprint," to gather innovative operating concepts and useful software from a wide range of organizations and individuals.

When Judge Chin scuttled the Google deal last year, Darnton got a historic opportunity to cast the DPLA as the world's best chance for a universal digital library. And indeed, it has gained broad support. Its plans have been praised by, among others, the Archivist of the United States, David Ferriero, and it has forged important partnerships, including one with Europeana, a European Commission–sponsored digital library with a similar concept.

At the same time, the DPLA's decision to call itself a "public library" has raised hackles. At a meeting in May of 2012, a group called the Chief Officers of State Library Agencies passed a resolution asking the DPLA steering committee to change the name of the project. While the state librarians expressed support for an effort to "make the cultural and scientific heritage of our country and the world freely available to all," they worried that by presenting itself as the country's public library, the DPLA could lend credence to "the unfounded belief that public libraries can be replaced in over 16,000 communities in the U.S. by a national digital library." Such a perception would make it even harder for local libraries to protect their budgets from cuts. Other critics have seen arrogance in the DPLA's assumption that a single online library can support the very different needs of scholarly researchers and the public. To strengthen its ties to public libraries, the DPLA added five public librarians to its steering committee last year, including Boston Public Library president Amy Ryan and San Francisco city librarian Luis Herrera.

The blowup over the name points to a deeper problem confronting the nascent online library: its inability to define itself. The DPLA remains a mystery in many ways. No one knows precisely how it will operate or even what exactly it will be. Some of the vagueness is deliberate. When the Berkman Center launched the initiative, it wanted major decisions to be made in a collaborative and inclusive manner, avoiding top-down decrees that might alienate any of its many constituencies. But according to current DPLA officials and others involved in the project, the seventeen members of the steering committee also have fundamental disagreements about the library's mission and scope. Many important aspects of the effort remain, in Palfrey's words, "to be determined."

No consensus has been reached, for example, on the extent to which the DPLA will host digitized books on its own servers, as opposed to providing links to digital collections stored on the computers of other libraries and archives. Nor has the steering committee made a firm decision about which materials other than books will be included in the library. Photographs, motion pictures, audio recordings, images of objects, and even blog posts and online videos are all under consideration. Another open question, one with particularly far-reaching implications, is whether the DPLA will try to provide any sort of access to recently published books, including popular ebooks. Darnton, for his part, believes that the digital library should steer clear of works published in the last five or ten years, to avoid treading on the turf of publishers and public libraries. It would be a mistake, he warns, for the DPLA to "invade the current commercial market." But while he says he has yet to hear anyone make a convincing counterargument, he admits that his view may not be held by everyone. Palfrey will only say that the DPLA is studying the issue of ebook lending but has yet to decide whether its scope will extend to recent publications.

Also unsettled is the critical question of how the DPLA will present itself to the public. David Weinberger, a Berkman researcher who is overseeing the development of the library's technical platform, says that no decision has been reached on whether the DPLA will offer a "front-end interface," such as a website or a smartphone app, or restrict itself to being a behind-the-scenes data clearinghouse that other organizations can tap into. The technology team's immediate goals are relatively modest. First the group wants to establish a flexible, open-source protocol for importing catalog information, borrowing statistics, and other data from participating institutions. Then it aims to organize that metadata into a unified database. And next it wants to provide an open programming interface for the database, with the hope of inspiring creative coders to develop useful applications. Palfrey says that he expects the DPLA to operate its own public website, but he is wary of making any predictions about the functions of that site or the degree to which it may overlap with the online offerings of traditional libraries. While he hopes that the DPLA will be more than a "metadata repository," he also says he would consider the effort a success even if it ultimately provided just the "plumbing" required to connect diverse and far-flung collections of materials.

It's hardly surprising that a large and diverse steering committee would have difficulty reaching unanimity on complicated and controversial matters. And it's understandable that the DPLA's leaders would be nervous about making concrete decisions that would almost certainly upset some people in the library profession and the publishing business. But there's growing tension between the heroic self-portrait that the DPLA presents to the public—its website proclaims that it "will make the cultural and scientific heritage of humanity available, free of charge, to all"—and the tentativeness and equivocation that cloud what is actually being built. If the uncertainties about the

DPLA's identity and workings aren't cleared up, they could end up delaying or even waylaying the project.

EVEN IF the views of the steering committee members were to come into harmony tomorrow, the ultimate form of the DPLA would remain hazy. The biggest question hanging over the project is one that can't be decided by executive fiat, or even by methodical consensus building. It's the same question that confronted Google Book Search and that bedevils every other effort to create an expansive online library: How do you navigate the country's onerous copyright restrictions? "The legal problems are staggering," Darnton says.

The U.S. Congress passed the first federal copyright law in 1790. Following English precedent, lawmakers sought to strike a reasonable balance between the desire of writers to earn a living and the benefit to society of giving people free access to the ideas of others. The law allowed "Authors and Proprietors" of "Maps, Charts and Books" to register a copyright in their work for fourteen years and, if they were still alive at the end of that term, to renew the copyright for another fourteen years. By limiting copy protections to a maximum of twenty-eight years, the legislators guaranteed that no book would remain under private control for very long. And by requiring that copyrights be formally registered, they ensured that most works would immediately enter the public domain. Of the thirteen thousand books published in the country during the decade following the law's enactment, fewer than six hundred were registered for copyright, according to historian John Tebbel.

Beginning in the 1970s, Congress developed a very different approach. Under pressure from film studios and other media and entertainment companies, legislators passed a series of bills that dramatically lengthened the term of copyright, not only for new books but retroactively for books published throughout most of the last century. Today, copyright in a work extends seventy years beyond the date of the author's death. Congress also removed the requirement that an author register a copyright—and, again, it applied the change retro-actively. Now a copyright is established for any work the moment it's created. Even when writers have no interest in claiming a copyright, they get one—and their works remain out of the public domain for decades. The upshot is that most books or articles written since 1923 remain off limits for unauthorized copying and distribution. Other nations have enacted similar policies, as part of an effort to establish international standards for trade in intellectual property.

Politicians make lousy futurists. As Google and the DPLA can tes-tify, the copyright changes put severe constraints on any attempt to scan, store, and provide online access to books published during most of the last hundred years. Moreover, the removal of the registration requirement means that millions of so-called orphan books—ones whose copyright holders either are unknown or can't be found—now lie beyond the reach of online libraries. Copyright protections are vitally important to ensuring that writers and artists have the wherewithal to create their works. But it's hard to look at the current situation without concluding that the restrictions have become so broad as to hamper the very creativity they were supposed to encourage. "Innovation is often being restricted today for legal reasons, not technological ones," says David K. Levine, an economist at Washington University in St. Louis and coauthor of *Against Intellectual Monopoly*. In many areas, he says, "people aren't creating new products because they fear a night-mare of copyright litigation."

There's a further twist. Books and other creative works behind the copyright wall aren't all that could be off-limits. Much of the metadata that libraries employ to catalog their holdings falls into a gray area with regard to how it can be reused. That's because many libraries purchase or license metadata from commercial suppliers or from the OCLC, a large library cooperative that syndicates an array of catalog-ing information. And because librarians have long used metadata from many sources in classifying their holdings, it can be extraordinarily difficult to sort out what's under license and what's not, or who owns what rights. The confusion makes even the DPLA's seemingly modest effort to collect metadata fraught with complications, according to David Weinberger. He says the DPLA is making progress at solving this problem, but when the library opens its virtual doors, patrons may have to make do with scanty descriptions of its contents.

SOME SCHOLARS believe that copyright restrictions will frustrate any attempt to create a universal online library unless Congress changes the law. James Grimmelmann, a copyright expert at New York Law School, feels that it will be "very, very hard" to include orphan works in a digital database without new legislation. Siva Vaidhyanathan, a University of Virginia media studies professor who hopes to start an international project to organize research materials online, believes that major changes in copyright law are essential to creating a digital library that includes recent works. He senses that it may take many years of public pressure to get politicians to deliver the necessary remedies.

While Palfrey is hesitant to discuss legal issues, he expresses some hope that progress can be made without congressional action. He feels that the DPLA may be able to hash out an agreement with publishers and authors that would enable it to offer copies of at least some of the orphans and other books published since 1923. Because of its nonprofit status, the DPLA may, according to some copyright experts, have an advantage over Google Book Search in negotiating such an agreement and getting it blessed by the courts.

The DPLA has made it clear that it will be meticulous in respecting copyrights. If it can't find a way around current legal constraints, whether through negotiation or legislation, it will have to limit its scope to books that are already in the public domain. And in that case, it's hard to see how it would be able to distinguish itself. After all, the web already offers plenty of sources for public-domain books. Google still provides full-text, searchable copies of millions of volumes published before 1923. So do the HathiTrust, a big book database run by a consortium of libraries, and Brewster Kahle's Internet Archive. Amazon's Kindle Store offers thousands of classic books free. And there's the venerable Project Gutenberg, which has been transcribing public-domain texts and putting them online since 1971 (when the project's founder, Michael Hart, typed the Declaration of Independence into a mainframe at the University of Illinois). Although the DPLA may be able to offer some valuable features of its own, including the ability to search collections of rare documents held by research libraries, those features would probably interest only a small group of scholars.

Despite the challenges it faces, the Digital Public Library of America has an enthusiastic corps of volunteers and some generous contributors. It seems likely that by this time next year, it will have reached its first milestone and begun operating a metadata exchange of some sort. But what happens after that? Will the library be able to extend the scope of its collection beyond the early years of the last century? Will it be able to offer services that spark the interest of the public? If the DPLA ends up nothing more than plumbing, the project will have failed to live up to its grand name and its even grander promise. The dream of H. G. Wells—and, for that matter, Robert Darnton—will have been deferred once again.

From MIT Technology Review
2012

THE BOYS OF
MOUNTAIN VIEW

IN DECEMBER OF 2001, an upstart Silicon Valley company named Google posted its corporate philosophy—in the form of a list of "ten things we have found to be true"—on its website. At once idealistic and smug, the list set the tone for the firm's future public pronouncements. "You can be serious without a suit" read one of the tenets. "You can make money without doing evil" went another. But it was the most innocuous-sounding of the commandments—"It's best to do one thing really, really well"—that would prove to be most fateful for the company. No sooner had Google pledged to remain a specialist than it broke its promise and began branching into new markets, with far-reaching consequences for its own business and the internet as a whole.

Google issued its philosophy at a decisive moment in its history. Although it had incorporated just three years earlier, in late 1998, its eponymous search engine was already widely viewed as the best tool available for navigating the net. But the company was struggling to make money. To succeed financially, its young founders, Stanford grad-school buddies Larry Page and Sergey Brin, knew they would have to supply not only search results but also advertisements tied to those results. At the time, the market for search-linked ads was dominated by another internet startup, Overture, which had forged partnerships with major web portals like Yahoo and America Online. Google's own advertising system, AdWords, was more sophisticated than Overture's, but big websites feared that the company, which operated its own site at Google.com, might end up competing with them for traffic. Google's high-toned philosophy, with its promise to stick to doing "one thing"—that is, web search—"really, really well," was meant to reassure would-be partners that it wouldn't raid their markets. The subtext was clear: "You can trust us; we're pure."

The tactic worked. During the course of 2002, Google signed advertising deals with several portals and other leading sites, including Earthlink, Ask Jeeves, and, most important, America Online. It vanquished Overture, the remains of which would be acquired by Yahoo, and it set itself on course to becoming the profit machine it is today. Things didn't work out quite so well for Google's partners, many of which would come to rue the role they played in securing the fledgling company's success. As Google grew, it expanded well beyond its initial focus on search, jumping into email, news aggregation, instant messaging, maps, financial advice, and video distribution, among many, many other businesses. Although the phrase "It's best to do one thing really, really well" remained a part of Google's official platform, by the end of 2005 the company had added an asterisk to the plank. "Over time," the footnote coolly explained, "we've expanded our view of the range of services we can offer—web search, for instance, isn't the only way for people to access or use information—and products that [four years ago] seemed unlikely are now key aspects of our portfolio."

Google's zeal in launching new services has helped it gain a larger purchase on people's online lives and opened up lucrative new sources of advertising revenues. But that expansionary zeal may turn out to be the firm's Achilles' heel. Last November, the European Union opened an antitrust investigation into Google's business practices. In June, it was revealed that the U.S. Federal Trade Commission had launched a similar probe. The two investigations will try to determine whether Google uses its omnipresent search engine to funnel people to its other sites, to the detriment of competition and innovation. The question on the table is central to the way the internet runs: Can the navigator also be the destination?

THE MAN who wrote Google's ten-point philosophy, a marketer named Douglas Edwards, has written a memoir of his time at the company. *I'm Feeling Lucky: The Confessions of Google Employee Number 59* is not the most thorough or even-handed account of Google's formative years, but it is the first to be written by an insider. It's a breezy tale of what it's like to be on board an internet juggernaut when it takes off. But it also sheds light on the character of a company that has very quickly come to play an outsized role in commerce and culture, mediating people's connections with information and ideas in myriad ways.

Edwards joined Google at the end of 1999, just as the company's search engine was beginning to attract notice for the superiority of its results. He had the immense good fortune to be recruited when Page and Brin still offered generous stock grants to new hires. Although he's coy about the size of his eventual windfall, it's clear that when he left Google, not long after its IPO in 2004, he was a wealthy man. But Edwards was considerably less lucky in being hired as a marketing functionary at a company whose technology-obsessed founders openly and very vocally disdained marketing, viewing it as, at best, a necessary evil in the fallen world of big business. Early in Edwards's tenure, Sergey Brin suggests, quite seriously, that Google use its entire marketing budget "to inoculate Chechan refugees against cholera." When he's finally convinced that it may be both impractical and rash for Google to inject itself into a civil war in Russia, he offers another suggestion for the marketing dollars: "What if we gave out free Google-branded condoms to high-school students?"

Google's rise is breathtaking, and there's much to admire about the dedication, skill, and quirky aplomb of the company's founders, engineers, and scientists. Page and Brin, who seem to be forever at work, rollerblading through the tight corridors of the company's Mountain View offices at all hours of day and night, have no patience for the trappings of corporate bureaucracy. The only things that matter to them are the quality of a person's work and the incisiveness of his or her thinking. Every aspect of the company's organization, operations, and ethics is subjected to ferocious debate, and good ideas are embraced whether they come from a senior executive or the company cook. Originality and ingenuity are richly rewarded, and orthodoxy is viewed with suspicion, if not contempt. To call the company a meritocracy would be an understatement.

But as Edwards's tale progresses, it becomes clear that, for all Google's laudable qualities, the company is hampered by a narrow perspective and an insular culture. Its flaws, like its strengths, stem from the personalities of its founders. Page and Brin seem oblivious to the world beyond technology, taking little interest in art or policy making or even popular culture, and they have trouble understanding anything that can't be measured in precise units or otherwise expressed in numbers. When pushed outside their geeky comfort zone, they can be pigheaded and petulant. Brin, while obviously endowed with a sharp mind, comes off particularly poorly in Edwards's account. He flits through the story like a spoiled child. At one meeting, where difficult personnel decisions are being made, he shows up wearing a spandex cycling outfit and amuses himself by flying a remote-controlled toy aircraft over the heads of his colleagues. At another point he nixes a potential Google print ad because it includes pictures of what he terms "unattractive people." He explains: "Our ads should always be aesthetically pleasing so people will think happy thoughts when they think of Google."

Page is rarely so callous, but like his friend he seems blind to shades of gray, particularly when looking at his own company. Even his vocabulary is black and white. What's good is "Googley." What's bad is "Not Googley." Any outsider who dares to question Google's motives or criticize its actions is a "bastard." The word "evil" is tossed around carelessly. Such a blinkered and self-serving view of the world may be forgivable in a young entrepreneur trying to get an ambitious technology company off the ground, but as Google has grown and its influence expanded, its hubris has become a problem. It accounts for many of the missteps that have stained the company's reputation in recent years.

EDWARDS'S STORY ends when he leaves the company in March of 2005. In the years since, Google's power and influence have only grown. It has become the web's central clearinghouse for information and one of the internet's principal tollgates. As people spend more time and do more things online, they also perform more Google searches and click on more Google ads—and the business's coffers swell. But the company's recent history is not quite as buoyant as its bottom line suggests. While it has introduced several attractive products, such as the Android operating system for smartphones and the Google Apps suite of cloud-computing programs, it has failed to discover strong new sources of profit. Many of its most hyped services—Google Base, Google Wave, Google Buzz, Google Health—have fizzled. Designed by engineers, they proved too complicated for mere mortals. Its ambitious Google Books initiative has run into a wall of litigation, due in part to Page's arrogance in rushing to scan copyrighted books without considering their owners' interests.

The company has also come in for ethical criticism, which must be particularly irksome to its high-minded founders. When Google disclosed that it was censoring results for searches in China, in response to pressure from the government, it came in for withering criticism. It later reversed course and shut down its Chinese search engine. It has also displayed a cavalier attitude toward cultural sensitivities, particularly when it comes to safeguarding personal privacy. When it sent camera-equipped cars down streets in Europe, to assemble a database of images for its Street View mapping service, the move sparked widespread public protests and, in Germany, a criminal inquiry. There are companies with worse records than Google when it comes to exploiting personal data and kowtowing to authoritarian regimes. But when you've publicly pledged to "make money without doing evil," you shouldn't be surprised to be held to a high standard.

Google is now trying to mend some of its overseas rifts with a time-tested strategy: using big investments to curry favor with local authorities. As the *New York Times* reported earlier this year, it has sent top executives to European countries to "dispense chunks of the company's $36 billion in cash reserves." In Ireland, which is struggling to emerge from an economic slump, Google bought a large Dublin office building from a government agency charged with cleaning up bad real estate loans. In France, it is setting up a European cultural center in its new Paris headquarters, the former home of the insurance giant AXA. In Germany, it is funding the establishment of an academic institute in Berlin that will study online privacy and other internet issues. Throughout the job-strapped continent, it is ratcheting up its hiring. It plans to add a thousand European employees this year alone.

The largest and most ominous challenge facing the company comes from its own backyard. The recent rise of vast, self-contained social networks is changing the way people use the net, making them less reliant on traditional search engines for finding information. Ambitious, youthful companies like Facebook and Twitter have begun to make Google look like a wallflower. While the company's brilliance at crunching numbers and building elaborate computer systems remains unquestioned, its lack of social skills threatens to become a major competitive liability.

The influx of new competitors also makes the timing of the new antitrust investigations seem a little ironic. Some observers argue that the FTC's probe may in fact be a signal that Google's influence has already peaked. They point out that the United States filed its antitrust suit against Microsoft in 1998, the very moment that power in the computer business began ebbing away from the PC giant and toward innovative internet companies, like Google. It's important to recognize, though, that Google's worldwide control over web search remains overwhelming, and despite its recent stumbles it continues to roll out new services intended to head off competitors and pull web users back into its fold. If, as critics allege, Google tweaks its search results to give undue precedence to its own sites and services, it could sway the future of many online markets, to its own benefit.

In April, Larry Page was named Google's chief executive. He replaced Eric Schmidt, a tech-industry veteran who had been hired years earlier to provide Page and Brin with, as Schmidt himself put it, "adult supervision." Page takes command at another crucial moment for Google—and the internet. The way he responds to the antitrust investigations, to the company's aggressive new rivals, and to persistent public concerns about online privacy and security will determine whether Google flourishes or flounders in the years ahead. His success will likely hinge on his ability to get beyond a black-and-white, us-versus-them view of the challenges facing his company, to realize that even bastards may have a point. He'll have to become a little less Googley and a little more worldly.

That won't be easy. Edwards begins his book with an anecdote about a meeting he had with Page back in 2002. Bruised by the founder's tendency to dismiss or ignore his suggestions, the marketer arrives at Page's office looking to ingratiate himself with his prickly boss. "I know I haven't always agreed with the direction you and Sergey have set for us," he says. "But I've been thinking about it and I just want to tell you that, in looking back, I realize that more often than not you've been right about things." Page looks up from his computer screen, a befuddled expression on his face. "More often than not?" he replies. "When were we *ever* wrong?"

From The National Interest
2011

THE EUNUCH'S CHILDREN

GUTENBERG WE KNOW. But what of the eunuch Cai Lun?

A well-educated, studious young man, a close aide to the Emperor Hedi in the Chinese imperial court of the Eastern Han Dynasty, Cai invented paper one fateful day in the year 105. At the time, writing and drawing were done primarily on silk, which was elegant but expensive, or on bamboo, which was sturdy but cumbersome. Seeking a more practical alternative, Cai came up with the idea of mashing bits of tree bark and hemp fiber together in a little water, pounding the resulting paste flat with a stone mortar, and then letting it dry into sheets in the sun. The experiment was a success. Allowing for a few industrial tweaks, Cai's method is still pretty much the way paper gets made today.

Cai killed himself some years later, having become entangled in a palace scandal from which he saw no exit. But his invention took on a life of its own. The craft of papermaking spread quickly throughout China and then, following the Silk Road westward, made its way into Persia, Arabia, and Europe. Within a few centuries, paper had replaced animal skins, papyrus mats, and wooden tablets as the world's preferred medium for writing and reading. The goldsmith Gutenberg would, with his creation of the printing press around 1450, mechanize the work of the scribe, replacing inky fingers with inky machines, but it was Cai Lun who gave us our reading material and, some would say, our world.

PAPER MAY be the single most versatile invention in history, its uses extending from the artistic to the bureaucratic to the hygienic. Rarely, though, do we give it its due. The ubiquity and disposability of the stuff—the average American goes through a quarter ton of it every year—lead us to take it for granted, or even to resent it. It's hard to respect something that you're forever throwing in the trash or flushing down the john or blowing your nose into. But modern life is inconceivable without paper. If paper were to disappear, writes Ian Sansom in his recent book *Paper: An Elegy*, "everything would be lost."

But wait. "An elegy"? Sansom's subtitle is half joking, but it's half serious, too. For while paper will be around as long as we're around, with the digital computer we have at last come up with an invention to rival Cai Lun's. Over the last decade, annual per-capita paper consumption in developed countries has fallen sharply. If the initial arrival of the personal computer and its companion printer had us tearing through more reams than ever, the rise of the internet as a universal communication system seems to be having the opposite effect. As more and more information comes to be stored and exchanged electronically, we're writing fewer checks, sending fewer letters, circulating fewer reports, and in general committing fewer thoughts to paper. Even our love notes are passed between servers.

In 1894, *Scribner's Magazine* published an essay by the French litterateur Octave Uzanne titled "The End of Books." Thomas Edison had recently invented the phonograph, and Uzanne thought it inevitable that books and periodicals would soon be replaced by "various devices for registering sound" that people would carry around with them. Flipping through printed sheets of paper demanded far too much effort from the modern "man of leisure," he argued. "Reading, as we practice it today, soon brings on great weariness; for not only does it require of the brain a sustained attention which consumes a large proportion of the cerebral phosphates, but it also forces our bodies into various fatiguing attitudes." The printing press and its quaint products were no match for modern technology.

You have to hand it to Uzanne. He anticipated the arrival of the audiobook, the iPod, and even the smartphone. About the obsolescence of the printed page, however, he was entirely wrong. Yet his prophesy would enjoy continuing popularity among the intelligentsia. It would come to be repeated over and over again during the twentieth century. Every time a new communication medium came along—radio, telephone, cinema, TV, CD-ROM—pundits would send out, usually in printed form, another death notice for the press. H. G. Wells wrote a book proclaiming that microfilm would replace the book.

In 2011, the Edinburgh International Book Festival featured a session titled—why mess with a winner?—"The End of Books." One of the participants, the Scottish novelist Ewan Morrison, declared that "within twenty-five years the digital revolution will bring about the end of paper books." Baby boomers, it seemed obvious to Morrison, would be the last generation to read words inked on pages. The future of the book and the magazine and the newspaper—the future of the word—lay in "e-publishing." The argument seemed entirely reasonable at the time. Unlike Uzanne, who was merely speculating, Morrison could point to hard facts about trends in reading and publishing. People were flocking to the screen. Paper was toast.

Now, just three years later, the picture has grown blurrier. There are new facts, equally hard, which suggest that words will continue to appear on sheets of paper for a good long while. Ebook sales, which skyrocketed after the launch of Amazon's Kindle in late 2007, have fallen back to earth in recent months, and sales of physical books have remained surprisingly resilient. Printed books still account for about three-quarters of overall book sales in the United States, and if sales of used books, which have been booming, are taken into account, that percentage probably rises even higher. A recent survey revealed that even the biggest fans of ebooks continue to purchase a lot of printed volumes.

Periodicals have had a harder go of it, thanks to the profusion of free alternatives online and the steep declines in print advertising. But subscriptions to print magazines seem to be stabilizing. Although some publications are struggling to survive, others are holding on to their readers. Digital subscriptions, while growing smartly, still represent only a tiny slice of the market, and a lot of magazine readers don't seem eager to switch to e-versions. A survey of owners of iPads and other tablet computers, conducted last year, found that three-quarters of them still prefer to read magazines on paper. There are even some glimmers in the beleaguered newspaper business. The spread of paywalls and the bundling of print and digital subscriptions appear to be tempering the long-term decline in print circulation. A few major papers have even gained some print readers of late.

What's striking is that the prospects for print have improved even as the use of media-friendly mobile computers and apps has exploded. If physical publications were dying, you would think their condition should be deteriorating rapidly now, not stabilizing.

OUR EYES tell us that the words and pictures on a screen are pretty much identical to the words and pictures on a piece of paper. But our eyes lie. What we're learning now is that reading is a bodily activity. We take in information the way we experience the world—as much with our sense of touch as with our sense of sight. Some scientists believe that our brain actually interprets written letters and words as physical objects, a reflection of the fact that our minds evolved to perceive things, not symbols of things.

The differences between page and screen go beyond the simple tactile pleasures of good paper stock. To the human mind, a sequence of pages bound together into a physical object is very different from a flat screen that displays only a single "page" of information at a time. The physical presence of the printed pages, and the ability to flip back and forth through them, turns out to be important to the mind's ability to navigate written works, particularly lengthy and complicated ones. Even though we don't realize it consciously, we quickly develop a mental map of the contents of a printed text, as if its argument or story were a voyage unfolding through space. If you've ever picked up a book you read long ago and discovered that your hands were able to locate a particular passage quickly, you've experienced this phenomenon. When we hold a physical publication in our hands, we also hold its contents in our mind.

The spatial memories seem to translate into more immersive reading and stronger comprehension. A recent experiment conducted with young readers in Norway found that, with both expository and narrative works, people who read from pages understand the text better than those who read the same material on a screen. The findings are consistent with a series of other recent reading studies. "We know from empirical and theoretical research that having a good spatial mental representation of the physical layout of the text supports reading comprehension," wrote the Norwegian researchers. They suggested that the ability of print readers to "see as well as tactilely feel the spatial extension and physical dimensions" of an entire text likely played a role in their superior comprehension.

That may also explain why surveys in the United States and other countries show that college students continue to prefer printed textbooks over electronic ones by wide margins. Students say that traditional books are more flexible as study tools, encourage deeper and more attentive reading, and promote better understanding and retention of the material. It seems to be true, as Octave Uzanne suggested, that reading printed publications consumes a lot of "cerebral phosphates." But maybe that's something to be celebrated.

Electronic books and periodicals have advantages of their own, of course. They're convenient. They often provide links to other relevant publications. Their contents can be searched and shared easily. They can include animations, audio and video snippets, and interactive features. They can be updated on the fly. When it comes to brief news reports or other simple stories, or works that we just want to glance at rather than read carefully, the screen may well be superior to the page.

We were probably mistaken to think of electronic publications as substitutes for printed ones. They seem to be different things, suited to different kinds of reading and providing different sorts of aesthetic and intellectual experiences. Some readers may continue to prefer print, others may develop a particular taste for the digital, and still others may happily switch back and forth between the two forms. This year in the United States, some two billion books and 350 million magazines will roll off presses and into people's hands. We are still Cai Lun's children.

From Nautilus
2013

PAST-TENSE POP

"Who wants yesterday's papers?" sang Mick Jagger in 1967. "Who wants yesterday's girl?" The answer, in the Swinging Sixties, was obvious: "Nobody in the world." That was then. Now we seem to want nothing more than to flip through yesterday's papers and carry on with yesterday's girl. Popular culture has become obsessed with the past—with recycling it, rehashing it, replaying it. Though we live in a fast-forward age, we can't take our finger off the Rewind button.

Nowhere is the past's grip so tight as in the world of music, as the rock critic Simon Reynolds makes clear in his new book *Retromania*. Over the last two decades, he argues, the "exploratory impulse" that once powered pop music forward has shifted its focus from Now to Then. Fans and musicians alike have turned into archeologists. The evidence is everywhere. There are the reunion tours and the reissues, the box sets and the tribute albums. There are the R&B museums, the rock halls of fame, the punk libraries. There are the collectors of vinyl and cassettes and—God help us—eight-tracks. There are the remixes, the samples, the "curated" playlists. When pop shakes its moneymaker today, what rises is the dust of the archive.

Nostalgia is nothing new. It's been a refrain of art and literature at least since Homer set Odysseus on Calypso's island and had him yearn to turn back time. And popular music has always had a strong revivalist streak, particularly in Reynolds's native Britain. But retromania is not just about nostalgia. It goes deeper than the tie-dyed dreams of Baby Boomers or the gray-flecked mohawks of Gen X punks. Whereas nostalgia is rooted in a sense of the past as past, retromania stems from a sense of the past as present. Yesterday's music, in all its forms, has become the atmosphere of contemporary culture. We live, Reynolds remarks, in "a simultaneity of pop time that abolishes history while nibbling away at the present's own sense of itself as an era with a distinct identity and feel."

One reason is the sheer quantity of pop music that has accumulated over the past half-century. Whether it's rock, funk, country, or electronica, we have heard it all before. Even the edgiest musicians have little choice but to produce pastiche. Greatly amplifying the effect is the recent shift to producing and distributing songs as digital files. When kids had to fork out cash for records or CDs, they had to make hard choices about what they listened to and what they let pass by. Usually, they would choose the new over the old, which served to keep the past at bay. Now, thanks to freely traded MP3s and all-you-can-eat music services such as Spotify, there's no need to make choices. Pretty much any song ever recorded is just a click away. With the economic barrier removed, the old floods in, swamping the new.

Reynolds argues that the glut of tunes has not just changed what we listen to; it has changed how we listen. The rapt fan who knew every hook, lyric, and lead by heart has been replaced by the fickle dabbler who cannot stop hitting Next. Reynolds presents himself as a case in point, and his experience will sound familiar to anyone with a hard drive packed with music files. He was initially "captivated" by the ability to use a computer to navigate an ocean of tunes. But in short order he found himself more interested in the mechanism than the music: "Soon I was listening to just the first fifteen seconds of every track; then, not listening at all." The logical culmination, he writes, "would have been for me to remove the headphones and just look at the track display."

Given a choice between more and less, we all choose more, even if it means a loss of sensory and emotional engagement. Though we don't like to admit it, the digital music revolution has just confirmed what we have always known: We cherish what is scarce, and what is abundant we view as disposable.

As all time is compressed into the present moment, our recycling becomes ever more compulsive. We begin to plunder not just bygone eras but the immediate past. Over the course of the last decade, writes

Reynolds, "the interval between something happening and its being revisited seemed to shrink insidiously." Not only did we have sixties revivals and seventies revivals and eighties revivals, but we even began to see revivals of musical fashions from the nineties, such as shoegaze and Britpop. It sometimes seems that the reason things go out of fashion so quickly these days is because we can't wait for them to come back into fashion. Displaying enthusiasm for something new is socially risky, particularly in an ironical time. It's safer to wait for it to come around again, preferably bearing the "vintage" label.

For musicians themselves, the danger is that their art becomes disconnected from the present—"timeless" in a bad sense. The eras of greatest ferment and creativity in popular music, such as the mid-sixties and the late seventies, were times of social discontent, when the young rejected the past and its stifling traditions. Providing the soundtrack for rebellion, rock musicians felt compelled to slay their fathers rather than pay tribute to them. Even if their lyrics were about getting laid or getting high—as they frequently were—their songs were filled with political force. Those not busy being born, as Dylan put it shortly after taking an axe to his folkie roots, are busy dying.

Now, youth culture is largely apolitical, and pop's soundtrack is just a soundtrack. Those not busy being born are busy listening to their iPods. Whether it's Fleet Foxes or Friendly Fires, Black Keys or Beach House, today's bands are less likely to battle the past than to luxuriate in it. That doesn't mean they aren't good bands. As Reynolds is careful to note, there is plenty of fine pop music being made today, in an ear-boggling array of styles. But drained of its subversive energies, none of it matters much. It just streams by.

Retromania is an important and often compelling work. It is also a sprawling one. Its aesthetic is more *Sandinista!* than "Hey Ya!" But Reynolds is sharp, and he knows his stuff. Even when his narrative wanders into the brush, the details remain sharp. (Who knew that the rave scene of the early nineties had its origins in the trad-jazz fad that preceded Beatlemania in England?) Reynolds might also be accused of being something of a retromaniac himself. After all, in worrying about the enervating influence of the past, he echoes the complaints of earlier cultural critics. "Our age is retrospective," grumbled Emerson in 1836. "Why should we grope among the dry bones of the past, or put the living generation into masquerade out of its faded wardrobe?" Longing for a less nostalgic time may itself be a form of nostalgia.

But Reynolds makes a convincing case that today's retromania is different in degree and in kind from anything we've experienced before. And it is not just an affliction of the mainstream. It has also warped the perspective of the avant-garde, dulling culture's cutting edge. It's one thing for old folks to look backward. It's another thing— and a far more lamentable one—for young people to feed on the past. Somebody needs to figure out a new way to smash a guitar.

From The New Republic
2011

THE LOVE THAT LAYS THE SWALE IN ROWS

THERE'S A LINE OF verse I'm always coming back to, and it's been on my mind more than usual these last few months:

The fact is the sweetest dream that labor knows.

It's the second to last line of one of Robert Frost's earliest and best poems, a sonnet called "Mowing." He wrote it just after the turn of the twentieth century, when he was a young man, in his twenties, with a young family. He was working as a farmer, raising chickens and tending a few apple trees on a small plot of land his grandfather had bought for him in Derry, New Hampshire. It was a difficult time in his life. He had little money and few prospects. He had dropped out of two colleges, Dartmouth and Harvard, without earning a degree. He had been unsuccessful in a succession of petty jobs. He was sickly. He had nightmares. His firstborn child, a son, had died of cholera at the age of three. His marriage was troubled. "Life was peremptory," Frost would later recall, "and threw me into confusion."

But it was during those lonely years in Derry that he came into his own as a writer and an artist. Something about farming—the long, repetitive days, the solitary work, the closeness to nature's beauty and carelessness—inspired him. The burden of labor eased the burden of life. "If I feel timeless and immortal it is from having lost track of time for five or six years there," he would write of his stay in Derry. "We gave up winding clocks. Our ideas got untimely from not taking newspapers for a long period. It couldn't have been more perfect if we had planned it or foreseen what we were getting into." In the breaks between chores on the farm, Frost somehow managed to write most of the poems for his first book, *A Boy's Will*; about half the poems for his second book, *North of Boston*; and a good number of other poems that would find their way into subsequent volumes.

"Mowing," from *A Boy's Will*, was the greatest of his Derry lyrics. It was the poem in which he found his distinctive voice: plainspoken and conversational, but also sly and dissembling. (To really understand Frost—to really understand anything, including yourself—requires as much mistrust as trust.) As with many of his best works, "Mowing" has an enigmatic, almost hallucinatory quality that belies the simple and homely picture it paints—in this case of a man cutting a field of grass for hay. The more you read the poem, the deeper and stranger it becomes:

> There was never a sound beside the wood but one,
> And that was my long scythe whispering to the ground.
> What was it it whispered? I knew not well myself;
> Perhaps it was something about the heat of the sun,
> Something, perhaps, about the lack of sound—
> And that was why it whispered and did not speak.
> It was no dream of the gift of idle hours,
> Or easy gold at the hand of fay or elf:
> Anything more than the truth would have seemed too weak
> To the earnest love that laid the swale in rows,
> Not without feeble-pointed spikes of flowers
> (Pale orchises), and scared a bright green snake.
> The fact is the sweetest dream that labor knows.
> My long scythe whispered and left the hay to make.

We rarely look to poetry for instruction anymore, but here we see how a poet's scrutiny of the world can be more subtle and discerning than a scientist's. Frost understood the essence of what we now call "embodied cognition" and the meaning of that heightened mental state we term "flow" long before psychologists and neurobiologists delivered the empirical evidence. His mower is not an airbrushed peasant, a rustic caricature. He's a farmer, a man doing a hard job on a still, hot summer day. He's not dreaming of "idle hours" or "easy gold." His mind is on his work—the bodily rhythm of the cutting, the weight of the tool in his hands, the stalks piling up around him. He's not seeking some greater truth beyond the work. The work is the truth.

The fact is the sweetest dream that labor knows.

There are mysteries in that line. Its power lies in its refusal to mean anything more or less than what it says. But it seems clear that what Frost is getting at, in the line and in the poem, is the centrality of action to both living and knowing. Only through work that brings us into the world do we approach a true understanding of existence, of "the fact." It's not an understanding that can be put into words. It can't be made explicit. It's nothing more than a whisper. To hear it, you need to get very near its source. Labor, whether of the body or the mind, is more than a way of getting things done. It's a form of contemplation, a way of seeing the world face-to-face rather than through a glass. Action un-mediates perception, gets us close to the thing itself. It binds us to the earth, Frost implies, as love binds us to one another. The antithesis of transcendence, work puts us in our place.

Frost is a poet of labor. He's always coming back to those revelatory moments when the active self blurs into the surrounding world—when, as he would write in another poem, "the work is play for mortal stakes." Richard Poirier, in his book *Robert Frost: The Work of Knowing*, described with great sensitivity the poet's view of the essence and essentialness of hard work: "Any intense labor enacted in his poetry, like mowing or apple-picking, can penetrate to the visions, dreams, myths that are at the heart of reality, constituting its articulate form for those who can read it with a requisite lack of certainty and an indifference to merely practical possessiveness." The knowledge gained through such efforts may be as shadowy and elusive as a dream, but "in its mythic propensities, the knowledge is less ephemeral than are the apparently more practical results of labor, like food or money." When we embark on a task, with our bodies or our minds, on our own or alongside others, we usually have a practical goal in sight. Our eyes are looking ahead to the product of our work—a store of hay for feeding livestock, perhaps. But it's through the work itself that we come to a deeper understanding of ourselves and our situation. The mowing, not the hay, is what matters most.

The Joy of Tools

Frost is not romanticizing some distant, pre-technological past. Although he was dismayed by those who allowed themselves to become "bigoted in reliance / On the gospel of modern science," he felt a kinship with scientists and inventors. As a poet, he shared with them a common spirit and pursuit. They were all explorers of the mysteries of earthly life, excavators of meaning from matter. They were all engaged in work that, as Poirier described it, "can extend the capability of human dreaming." For Frost, the greatest value of "the fact"—whether apprehended in the world or expressed in a work of art or made manifest in a tool or other invention—lay in its ability to expand the scope of individual knowing and hence open new avenues of perception, action, and imagination. In the long poem "Kitty Hawk," written near the end of his life, he celebrated the Wright brothers' flight "Into the unknown, / Into the sublime." In making their own "pass / At the infinite," the brothers also made the experience of flight, and the sense of unboundedness it provides, possible for all of us.

Technology is as crucial to the work of knowing as it is to the work of production. The human body, in its native, unadorned state, is a feeble thing. It's constrained in its strength, its dexterity, its sensory range, its calculative prowess, its memory. It quickly reaches the limits of what it can do. But the body encompasses a mind that can imagine, desire, and plan for achievements the body alone can't fulfill. This tension between what the body can accomplish and what the mind can envision is what gave rise to and continues to propel and shape technology. It's the spur for humankind's extension of itself and elaboration of nature. Technology isn't what makes us "posthuman" or "transhuman," as some writers and scholars these days suggest. It's what makes us human. Technology is in our nature. Through our tools we give our dreams form. We bring them into the world. The practicality of technology may distinguish it from art, but both spring from a similar, distinctly human yearning.

One of the many jobs the human body is unsuited to is cutting grass. (Try it if you don't believe me.) What allows the mower to do his work, what allows him to be a mower, is the tool he wields, his scythe. The mower is, and has to be, technologically enhanced. The tool makes the mower, and the mower's skill in using the tool remakes the world for him. The world becomes a place in which he can act as a mower, in which he can lay the swale in rows. This idea, which on the surface may sound trivial or even tautological, points to something elemental about life and the formation of the self.

"The body is our general means of having a world," wrote the French philosopher Maurice Merleau-Ponty in his 1945 masterwork *Phenomenology of Perception*. Our physical makeup—the fact that we walk upright on two legs at a certain height, that we have a pair of hands with opposable thumbs, that we have eyes which see in a particular way, that we have a certain tolerance for heat and cold—determines our perception of the world in a way that precedes, and then molds, our conscious thoughts about the world. We see mountains as lofty not because mountains are lofty but because our perception of their form and height is shaped by our own stature. We see a stone as, among other things, a weapon because the particular construction of our hand and arm enables us to pick it up and throw it. Perception, like cognition, is embodied.

It follows that whenever we gain a new talent, we not only change our bodily capacities, we change the world. The ocean extends an invitation to the swimmer that it withholds from the person who has never learned to swim. With every skill we master, the world reshapes itself to reveal greater possibilities. It becomes more interesting, and being in it becomes more rewarding. This may be what Baruch Spinoza, the seventeenth-century Dutch philosopher who rebelled against René Descartes' division of mind and body, was getting at when he wrote, "The human mind is capable of perceiving a great many things, and is the more capable, the more its body can be disposed in a great many ways." John Edward Huth, a physics professor at Harvard, testifies to the regeneration that attends the mastery of a skill. A decade ago, inspired by Inuit hunters and other experts in natural wayfinding, he undertook "a self-imposed program to learn navigation through environmental clues." Through months of rigorous outdoor observation and practice, he taught himself how to read the nighttime and daytime skies, interpret the movements of clouds and waves, decipher the shadows cast by trees. "After a year of this endeavor," he recalled in a recent essay, "something dawned on me: the way I viewed the world had palpably changed. The sun looked different, as did the stars." Huth's enriched perception of the environment, gained through a kind of "primal empiricism," struck him as being "akin to what people describe as spiritual awakenings."

Technology, by enabling us to act in ways that go beyond our bodily limits, also alters our perception of the world and what the world signifies to us. Technology's transformative power is most apparent in tools of discovery, from the microscope and the particle accelerator of the scientist to the canoe and the spaceship of the explorer, but the power is there in all tools, including the ones we use in our everyday lives. Whenever an instrument allows us to cultivate a new talent, the world becomes a different and more intriguing place, a setting of even greater opportunity. To the possibilities of nature are added the possibilities of culture. "Sometimes," wrote Merleau-Ponty, "the signification aimed at cannot be reached by the natural means of the body. We must, then, construct an instrument, and the body projects a cultural world around itself." The value of a well-made and well-used tool lies not only in what it produces for us but what it produces in us. At its best, technology opens fresh ground. It gives us a world that is at once more understandable to our senses and better suited to our intentions—a world in which we're more at home. Used thoughtfully and with skill, a tool becomes much more than a means of production or consumption. It becomes a means of experience. It gives us more ways to lead rich and engaged lives.

Look more closely at the scythe. It's a simple tool, but an ingenious one. Invented around 500 BC, by the Romans or the Gauls, it consists of a curved blade, forged of iron or steel, attached to the end of a long wooden pole, or snath. The snath typically has, about halfway down its length, a small wooden grip, or nib, that makes it possible to grasp and swing the implement with two hands. The scythe is a variation on the much older sickle, a similar but short-handled cutting tool that was invented in the Stone Age and came to play an essential role in the early development of agriculture and, in turn, of civilization. What made the scythe a momentous innovation in its own right is that its long snath allowed a farmer or other laborer to cut grass at ground level while standing upright. Hay or grain could be harvested, or a pasture cleared, more quickly than before. Agriculture leaped forward.

The scythe enhanced the productivity of the worker in the field, but its benefit went beyond what could be measured in yield. The scythe was a congenial tool, far better suited to the bodily work of mowing than the sickle had been. Rather than stooping or squatting, the farmer could walk with a natural gait and use both his hands, as well as the full strength of his torso, in his job. The scythe served as both an aid and an invitation to the skilled work it enabled. We see in its form a model for technology on a human scale, for tools that extend the productive capabilities of society without circumscribing the individual's scope of action and perception. Indeed, as Frost makes clear in "Mowing," the scythe intensifies its user's involvement with and apprehension of the world. The mower swinging a scythe does more, but he also knows more. Despite outward appearances, the scythe is a tool of the mind as well as the body.

Not all tools are so congenial. Some deter us from skilled action. The technologies of computerization and automation that hold such sway over us today rarely invite us into the world or encourage us to develop new talents that enlarge our perceptions and expand our possibilities. They mostly have the opposite effect. They're designed to be disinviting. They pull us away from the world. That's a consequence not only of prevailing design practices, which place ease and efficiency above all other concerns, but also of the fact that, in our personal lives, the computer has become a media device, its software painstakingly programmed to grab and hold our attention. As most people know from experience, the computer screen is intensely compelling, not only for the conveniences it offers but also for the many diversions it provides. There's always something going on, and we can join in at any moment with the slightest of effort. Yet the screen, for all its enticements and stimulations, is an environment of sparseness—fast-moving, efficient, clean, but revealing only a shadow of the world.

That's true even of the most meticulously crafted simulations of space that we find in virtual-reality applications such as games, architectural models, three-dimensional maps, and the tools used by surgeons and others to control robots. Artificial renderings of space may provide stimulation to our eyes and to a lesser degree our ears, but they tend to starve our other senses—touch, smell, taste—and greatly restrict the movements of our bodies. A study of rodents, published in *Science* in 2013, indicated that the brain cells used in navigation are much less active when animals make their way through computer-generated landscapes than when they traverse the real world. "Half of the neurons just shut up," reported one of the researchers, UCLA neurophysicist Mayank Mehta. He believes that the drop-off in mental activity likely stems from the lack of "proximal cues"—environmental smells, sounds, and textures that provide clues to location—in digital simulations of space. "A map is not the territory it represents," the Polish philosopher Alfred Korzybski famously remarked, and a computer rendering is not the territory it represents either. When we enter the virtual world, we're required to shed much of our body. That doesn't free us; it emaciates us.

The world in turn is made less meaningful. As we adapt to our streamlined environment, we render ourselves incapable of perceiving what the world offers its most ardent inhabitants. Like drivers following GPS commands, we travel blindfolded. The result is existential impoverishment, as nature and culture withdraw their invitations to act and to perceive. The self can only thrive, can only grow, when it encounters and overcomes "resistance from surroundings," wrote the American pragmatist John Dewey in *Art as Experience.* "An environment that was always and everywhere congenial to the straightaway execution of our impulses would set a term to growth as sure as one always hostile would irritate and destroy. Impulsion forever boosted on its forward way would run its course thoughtless, and dead to emotion."

Ours may be a time of material comfort and technological wonder, but it's also a time of aimlessness and gloom. During the first decade of this century, the number of Americans taking prescription drugs to treat depression or anxiety rose by nearly a quarter. One in five adults now regularly takes such medications. The suicide rate among middle-age Americans increased by nearly 30 percent over the same ten years, according to a report from the Centers for Disease Control and Prevention. More than 10 percent of American schoolchildren, and nearly 20 percent of high school–age boys, have been given a diagnosis of attention deficit hyperactivity disorder, and two-thirds of that group take drugs like Ritalin and Adderall to treat the condition. The reasons for our discontent are many and only dimly understood. But one of them may be that through the pursuit of a frictionless existence, we've succeeded in turning the landscape of our lives into a barren place. Drugs that numb the nervous system provide a way to rein in our vital, animal sensorium, to shrink our being to a size that better suits our constricted environs.

Neither Master nor Slave

Frost's sonnet also contains, as one of its many whispers, a warning about technology's ethical hazards. There's a brutality to the mower's scythe. It indiscriminately cuts down flowers—those tender, pale

orchises—along with the stalks of grass.* It frightens innocent animals, like the bright green snake. If technology embodies our dreams,
it also embodies other, less benign qualities in our makeup, such as
our will to power and the arrogance and insensitivity that accompany
it. Frost returns to this theme a little later in *A Boy's Will*, in a second
lyric about cutting hay, "The Tuft of Flowers." The poem's narrator
comes upon a freshly mown field and, while following the flight of a
passing butterfly with his eyes, discovers in the midst of the cut grass
a small cluster of flowers, "a leaping tongue of bloom" that "the scythe
had spared":

The mower in the dew had loved them thus,
By leaving them to flourish, not for us,

Nor yet to draw one thought of us to him,
But from sheer morning gladness to the brim.

Working with a tool is never just a practical matter, Frost is telling
us, with characteristic delicacy. It always entails moral choices and has
moral consequences. It's up to us, as users and makers of tools, to
humanize technology, to aim its cold blade wisely. That requires vigi-
lance and care.

The scythe is still employed in subsistence farming in many parts of
the world. But it has no place on the modern farm, the development of
which, like the development of the modern factory, office, and home,
has required ever more complex and efficient equipment. The threshing
machine was invented in the 1780s, the mechanical reaper appeared
around 1835, the baler came a few years after that, and the combine
harvester began to be produced commercially toward the end of the
nineteenth century. The pace of technological advance has only accel-

---

*The destructive potential of the scythe gains greater symbolic resonance when one
remembers that the orchis, a tuberous plant, derives its name from the Greek word
for testicle, *orkhis*. Frost was well versed in classical languages and literature.

erated in the decades since, and today the trend is reaching its logical conclusion with the computerization of agriculture. The working of the soil, which Thomas Jefferson saw as the most vigorous and virtuous of occupations, is being off-loaded almost entirely to machines. Farmhands are being replaced by "drone tractors" and other robotic systems that, using sensors, satellite signals, and software, plant seeds, fertilize and weed fields, harvest and package crops, and milk cows and tend other livestock. In development are robo-shepherds that guide flocks through pastures. Even if scythes still whispered in the fields of the industrial farm, no one would be around to hear them.

The congeniality of hand tools encourages us to take responsibility for their use. Because we sense the tools as extensions of our bodies, parts of ourselves, we have little choice but to be intimately involved in the ethical choices they present. The scythe doesn't choose to slash or spare the flowers; the mower does. As we become more expert in the use of a tool, our sense of responsibility for it naturally strengthens. To the novice mower, a scythe may feel like a foreign object in the hands; to the accomplished mower, hands and scythe become one thing. Talent tightens the bond between an instrument and its user. This feeling of physical and ethical entanglement doesn't have to go away as technologies become more complex. In reporting on his historic solo flight across the Atlantic in 1927, Charles Lindbergh spoke of his plane and himself as if they were a single being: "*We* have made this flight across the ocean, not *I* or *it*." The airplane was a complicated system encompassing many components, but to a skilled pilot it still had the intimate quality of a hand tool. The love that lays the swale in rows is also the love that parts the clouds for the stick-and-rudder man.

Automation weakens the bond between tool and user not because computer-controlled systems are complex but because they ask so little of us. They hide their workings in secret code. They resist any involvement of the operator beyond the bare minimum. They discourage the development of skillfulness in their use. Automation ends up having an anesthetizing effect. We no longer feel our tools as parts of ourselves. In a renowned 1960 paper, "Man-Computer Symbiosis," the psychologist and engineer J. C. R. Licklider described the shift in our relation to technology well. "In the man-machine systems of the past," he wrote, "the human operator supplied the initiative, the direction, the integration, and the criterion. The mechanical parts of the systems were mere extensions, first of the human arm, then of the human eye." The introduction of the computer changed all that. " 'Mechanical extension' has given way to replacement of men, to automation, and the men who remain are there more to help than to be helped." The more automated everything gets, the easier it becomes to see technology as a kind of implacable, alien force that lies beyond our control and influence. Attempting to alter the path of its development seems futile. We press the on switch and follow the programmed routine.

To adopt such a submissive posture, however understandable it may be, is to shirk our responsibility for managing progress. A robotic harvesting machine may have no one in the driver's seat, but it is every bit as much a product of conscious human thought as a humble scythe is. We may not feel the machine as part of our body, as we do the hand tool, but on an ethical level the machine still operates as an extension of our will. Its intentions are our intentions. If a robot scares a bright green snake (or worse), we're still to blame. We shirk a deeper responsibility as well: that of overseeing the conditions for the construction of the self. As computer systems and software applications come to play an ever larger role in shaping our lives and the world, we have an obligation to be more, not less, involved in decisions about their design and use—before the forward press of progress forecloses our options. We should be careful about what we make.

If that sounds naive or hopeless, it's because we have been misled by a metaphor. We've defined our relation with technology not as that of body and limb or even that of sibling and sibling but as that of master and slave. The idea goes way back. It took hold at the dawn of Western philosophical thought, emerging first with the ancient Athenians. Aristotle, in discussing the operation of households at the beginning of his *Politics*, argued that slaves and tools are essentially equivalent, the former acting as "animate instruments" and the latter as "inanimate instruments" in the service of the master of the house. If tools could somehow become animate, Aristotle posited, they would be able to substitute directly for the labor of slaves. "There is only one condition on which we can imagine managers not needing subordinates, and masters not needing slaves," he mused, anticipating the arrival of computer automation and even machine learning. "This condition would be that each [inanimate] instrument could do its own work, at the word of command or by intelligent anticipation." It would be "as if a shuttle should weave itself, and a plectrum should do its own harp-playing."

The conception of tools as slaves has colored our thinking ever since. It informs society's recurring dream of emancipation from toil. "All unintellectual labour, all monotonous, dull labour, all labour that deals with dreadful things, and involves unpleasant conditions, must be done by machinery," wrote Oscar Wilde in 1891. "On mechanical slavery, on the slavery of the machine, the future of the world depends." John Maynard Keynes, in a 1930 essay, predicted that mechanical slaves would free humankind from "the struggle for subsistence" and propel us to "our destination of economic bliss." In 2013, *Mother Jones* columnist Kevin Drum declared that "a robotic paradise of leisure and contemplation eventually awaits us." By 2040, he forecast, our computer slaves—"they never get tired, they're never ill-tempered, they never make mistakes"—will have rescued us from labor and delivered us into a new Eden. "Our days are spent however we please, perhaps in study, perhaps playing video games. It's up to us."

With its roles reversed, the metaphor also informs society's nightmares about technology. As we become dependent on our technological slaves, the thinking goes, we turn into slaves ourselves. From the eighteenth century on, social critics have routinely portrayed factory machinery as forcing workers into bondage. "Masses of labourers," wrote Marx and Engels in their *Communist Manifesto*, "are daily and hourly enslaved by the machine." Today, people complain all the time about feeling like slaves to their appliances and gadgets. "Smart devices are sometimes empowering," observed *The Economist* in "Slaves to the

Smartphone," an article published in 2012. "But for most people the servant has become the master." More dramatically still, the idea of a robot uprising, in which computers with artificial intelligence transform themselves from our slaves to our masters, has for a century been a central theme in dystopian fantasies about the future. The very word "robot," coined by a science fiction writer in 1920, comes from *robota*, a Czech term for servitude.

The master-slave metaphor, in addition to being morally fraught, distorts the way we look at technology. It reinforces the sense that our tools are separate from ourselves, that our instruments have an agency independent of our own. We start to judge our technologies not on what they enable us to do but rather on their intrinsic qualities as products— their cleverness, their efficiency, their novelty, their style. We choose a tool because it's new or it's cool or it's fast, not because it brings us more fully into the world and expands the ground of our experiences and perceptions. We become mere consumers of technology.

The metaphor encourages society to take a simplistic and fatalistic view of technology and progress. If we assume that our tools act as slaves on our behalf, always working in our best interest, then any attempt to place limits on technology becomes hard to defend. Each advance grants us greater freedom and takes us a stride closer to, if not utopia, then at least the best of all possible worlds. Any misstep, we tell ourselves, will be quickly corrected by subsequent innovations. If we just let progress do its thing, it will find remedies for the problems it creates. "Technology is not neutral but serves as an overwhelming positive force in human culture," writes one pundit, expressing the self-serving Silicon Valley ideology that in recent years has gained wide currency. "We have a moral obligation to increase technology because it increases opportunities." The sense of moral obligation strengthens with the advance of automation, which, after all, provides us with the most animate of instruments, the slaves that, as Aristotle anticipated, are most capable of releasing us from our labors.

The belief in technology as a benevolent, self-healing, autonomous force is seductive. It allows us to feel optimistic about the future while relieving us of responsibility for that future. It particularly suits the interests of those who have become extraordinarily wealthy through the labor-saving, profit-concentrating effects of automated systems and the computers that control them. It provides our new plutocrats with a heroic narrative in which they play starring roles: Recent job losses may be unfortunate, but they're a necessary evil on the path to the human race's eventual emancipation by the computerized slaves that our benevolent enterprises are creating. Peter Thiel, a successful entrepreneur and investor who has become one of Silicon Valley's most prominent thinkers, grants that "a robotics revolution would basically have the effect of people losing their jobs." But, he hastens to add, "it would have the benefit of freeing people up to do many other things." Being freed up sounds a lot more pleasant than being fired.

There's a callousness to such grandiose futurism. As history reminds us, high-flown rhetoric about using technology to liberate workers often masks a contempt for labor. It strains credulity to imagine today's technology moguls, with their libertarian leanings and impatience with government, agreeing to the kind of vast wealth-redistribution scheme that would be necessary to fund the self-actualizing leisure-time pursuits of the jobless multitudes. Even if society were to come up with some magic spell, or magic algorithm, for equitably parceling out the spoils of automation, there's good reason to doubt whether anything resembling the "economic bliss" imagined by Keynes would ensue.

In a prescient passage in her 1958 book *The Human Condition*, Hannah Arendt observed that if automation's utopian promise were actually to pan out, the result would probably feel less like paradise than like a cruel practical joke. The whole of modern society, she wrote, has been organized as "a laboring society," where working for pay, and then spending that pay, is the way people define themselves and measure their worth. Most of the "higher and more meaningful activities" revered in the distant past have been pushed aside or forgotten, and "only solitary individuals are left who consider what they are doing in terms of work and not in terms of making a living." For technology to fulfill humankind's abiding "wish to be liberated from labor's 'toil and trouble'" at this point would be perverse. It would cast us deeper into a purgatory of malaise. What automation confronts us with, Arendt concluded, "is the prospect of a society of laborers without labor, that is, without the only activity left to them. Surely, nothing could be worse." Utopianism, she understood, is a form of self-delusion.

A Question

A few months ago, I had a chance meeting on the campus of a small college with a freelance photographer who was working on an assignment for the school. He was standing under a tree, waiting for some uncooperative clouds to get out of the way of the sun. I noticed he had a large-format film camera set up on a bulky tripod—it was hard to miss, as it looked almost absurdly old-fashioned—and I asked him why he was still using film. He told me that he had eagerly embraced digital photography a few years earlier. He had replaced his film cameras and his darkroom with digital cameras and a computer running the latest image-processing software. But after a few months, he switched back. It wasn't that he was dissatisfied with the operation of the equipment or the resolution or accuracy of the images. It was that the way he went about his work had changed.

The constraints inherent in taking and developing pictures on film—the expense, the toil, the uncertainty—had encouraged him to work slowly when he was on a shoot, with deliberation, thoughtfulness, and a deep, physical sense of presence. Before he took a picture, he would compose the shot in his mind, attending to the scene's light, color, framing, and form. He would wait patiently for the right moment to release the shutter. With a digital camera, he could work faster. He could take a slew of images, one after the other, and then use his computer to sort through them and crop and tweak the most promising ones. The act of composition took place after a photo was taken. The change felt intoxicating at first. But he found himself disappointed with the results. The images left him cold. Film, he realized, imposed a discipline of perception, of seeing, which led to richer, more artful, more moving photographs. Film demanded more of him. And so he went back to the older technology.

The photographer wasn't the least bit antagonistic toward computers. He wasn't beset by any abstract concerns about a loss of agency or autonomy. He wasn't a crusader. He just wanted the best tool for the job—the tool that would encourage and enable him to do his finest, most fulfilling work. What he came to realize is that the newest, most automated, most expedient tool is not always the best choice. Although I'm sure he would bristle at being likened to the Luddites, his decision to forgo the latest technology, at least in some stages of his work, was an act of rebellion resembling that of the old English machine-breakers, if without the fury. Like the Luddites, he understood that decisions about technology are also decisions about ways of working and ways of living—and he took control of those decisions rather than ceding them to others or giving way to the momentum of progress. He stepped back and thought critically about technology.

As a society, we've become suspicious of such acts. Out of ignorance or laziness or timidity, we've turned the Luddites into cartoon characters, emblems of backwardness. We assume that anyone who rejects a new tool in favor of an older one is guilty of nostalgia, of making choices sentimentally rather than rationally. But the real sentimental fallacy is the assumption that the new thing is always better suited to our purposes and intentions than the old thing. That's the view of a child, naive and pliable. What makes one tool superior to another has nothing to do with how new it is. What matters is how it enlarges us or diminishes us, how it shapes our experience of nature and culture and one another. To cede choices about the texture of our daily lives to a grand abstraction called progress is folly.

Technology is a pillar and a glory of civilization. But it is also a test that we set for ourselves. It challenges us to think about what's important in our lives, to ask ourselves what *human being* means. Computerization, as it extends its reach into the deepest spheres of our existence, raises the stakes. We can allow ourselves to be carried along by the technological current, wherever it may be taking us, or we can push against it. To resist invention is not to reject invention. It's to humble invention, to bring progress down to earth. "Resistance is futile," goes the glib *Star Trek* cliché beloved by techies. But that's the opposite of the truth. Resistance is never futile. If the source of our vitality is, as Emerson taught us, "the active soul," then our highest obligation is to resist any force, whether institutional or commercial or technological, that would enfeeble or enervate the soul.

One of the most remarkable things about us is also one of the easiest to overlook: each time we collide with the real, we deepen our understanding of the world and become more fully a part of it. While we're wrestling with a challenge, we may be motivated by an anticipation of the ends of our labor, but, as Frost saw, it's the work—the means—that makes us who we are. Automation severs ends from means. It makes getting what we want easier, but it distances us from the work of knowing. As we transform ourselves into creatures of the screen, we face an existential question: Does our essence still lie in what we know, or are we now content to be defined by what we want?

From The Glass Cage
2014

THE SNAPCHAT CANDIDATE

Barack Obama killed it on social media this summer. On August 14, a Friday, he kicked off a steamy Washington weekend by releasing a pair of playlists, one for the nighttime and one for the day, through the White House's new Spotify account. The presidential mixes were predictable but pleasant, smooth fusions of dad rock and dad soul. In an accompanying blog post, one of the administration's digital functionaries promised that more playlists were in the works, including "issue-specific" ones.

Two weeks later, on the evening of August 31, Obama turned himself into the country's Instagrammer-in-Chief. While en route to Alaska to promote his climate agenda, the president took a photograph of a mountain range from a window on Air Force One and posted the shot on the popular picture-sharing network. "Hey everyone, it's Barack," the caption read. "I'll be spending the next few days touring this beautiful state and meeting with Alaskans about what's going on in their lives. Looking forward to sharing it with you." The photo was liked by thousands.

Ever since the so-called Facebook election of 2008, Obama has been a pacesetter in using social media to connect with the public. But he has nothing on this year's field of candidates. Back in June, the Hillary Clinton campaign issued its official Spotify playlist, packed with on-message tunes ("Brave," "Fighters," "Stronger," "Believer"). Ted Cruz live-streams his appearances on Periscope. Marco Rubio broadcasts "Snapchat Stories" along the trail. Rand Paul and Lindsey Graham produce goofy YouTube videos. Even grumpy old Bernie Sanders has attracted nearly two million friends on Facebook, leading the *New York Times* to dub him "a king of social media."

And then there's Donald Trump. If Sanders is a king, Trump is a god.

A natural-born troll, adept at issuing inflammatory bulletins at opportune moments, he's the first candidate optimized for the Google News algorithm. In a typical tweet, sent out at dawn at the start of a recent campaign week, he described Clinton aide Huma Abedin as "a major security risk" and "the wife of perv sleazebag Anthony Wiener." Exuberantly impolitic, such messages attract Trump a vast web audience—four million followers on Twitter alone—while giving reporters and pundits fresh bait to feed on. What Trump understands is that the best way to dominate the online discussion is not to inform but to provoke.

Trump's glow may fade—online celebrity has a fast-burning wick—but his ability to control the agenda this year says a lot about how social media is changing political discourse. We're learning that as the net shrinks to the size of a smartphone screen, the national conversation shrinks with it. The message has to fit the medium.

TWICE BEFORE in the last hundred years a new medium has transformed elections. In the 1920s, radio disembodied candidates, reducing them to voices. It also made national campaigns much more intimate. Politicians, used to bellowing at fairgrounds and train depots, found themselves talking to families in their homes. The blustery rhetoric that stirred big, partisan crowds came off as shrill and off-putting when piped into a living room or a kitchen. Gathered around their wireless sets, the public wanted an avuncular statesman, not a rabble-rouser. With Franklin Roosevelt, master of the soothing fireside chat, the new medium found its ideal messenger.

In the 1960s, television gave candidates their bodies back, at least in two dimensions. With its jumpy cuts and pitiless close-ups, TV placed a stress on sound bites, good teeth, and an easy manner. Image became everything, as the line between politician and celebrity blurred. John Kennedy was the first successful candidate of the TV era, but it was Ronald Reagan and Bill Clinton who perfected the form. Born actors, they managed to project a down-home demeanor while also seeming bigger than life. They were made for television.

Today, with the public looking to their phones for news and entertainment, we're at the start of the third technological makeover of modern electioneering. The presidential campaign is becoming just another social media stream, its swift and shallow current intertwining with all the other streams that flow through people's devices. This shift is changing the way politicians communicate with voters, altering the tone and content of political speech. But it's doing more than that. It's changing what the country wants and expects from its would-be leaders. If radio and TV required candidates to be nouns—to present themselves as stable, coherent figures—social media pushes them to be verbs, engines of activity. Authority and respect don't accumulate on social media; they have to be earned anew at each moment. You're only as relevant as your last tweet.

What's important now is not so much image as personality. But, as the Trump phenomenon suggests, it's only a particular kind of personality that works—one that's big enough to grab the attention of the perpetually distracted but small enough to fit neatly into a thousand tiny media containers. It might best be described as a Snapchat personality. It bursts into focus at regular intervals without ever demanding steady concentration.

Social media favors the bitty over the meaty, the cutting over the considered. It also prizes emotionalism over reason. The more visceral the message, the more quickly it circulates and the longer it holds the darting public eye. In something of a return to the pre-radio days, the fiery populist now seems more compelling, more worthy of attention, than the cool wonk. It's the crusty Bernie and the caustic Donald who get hearted and hash-tagged, friended and followed. Is it any wonder that "Feel the Bern" has become the rallying cry of the Sanders campaign?

Emotional appeals can be good for politics. They can spur civic involvement, even among the disenfranchised and disenchanted. And they can galvanize public attention, focusing it on injustices and abuses of power. An immediate emotional connection can, at best, deepen into a sustained engagement with the political process. But there's a dark side to social media's emotionalism. Trump's popularity took off only after he demonized Mexican immigrants, playing to the public's frustrations and fears. That's the demagogue's oldest tactic, and it worked. The Trump campaign may have qualities of farce, but it also suggests that a social media candidate, passionate yet hollow, could be a perfect vessel for a cult of personality.

WHENEVER A new medium upends the game, veteran politicians flounder. They go on playing by the old medium's rules. The people who listened to the 1960 Nixon-Kennedy debate on their radios were convinced Nixon had won. But the far larger television audience saw Kennedy as the clear victor. Nixon's mistake was to assume that he was still in the radio age. He thought the audience would concentrate on what he said and wouldn't care much about how he looked. Oblivious to the camera's gaze, he had no idea that the sweat on his upper lip was drowning out his words.

A similar inertia is hobbling the establishment candidates today. They continue to follow the conventions of TV elections. They assume that television will establish the campaign's talking points, package the race as a series of tidy stories, and shape the way voters see the contestants. The candidates may have teams of digital staffers tending to their online messaging, but most view social media as a complement to TV coverage, a means of reinforcing their messages and images, rather than a driving force in the race.

That's particularly true of Hillary Clinton and Jeb Bush, the erstwhile shoo-ins. Both are playing it safe online, trying to burnish their images as reliable public servants while avoiding any misstep that might blow up into a TV controversy. Bush's various social media feeds come off as afterthoughts. They promote his appearances, offer kudos to his endorsers, and provide links to his merchandise store. What they rarely do is break news. Clinton's postings are equally anodyne. Her Facebook feed is a mirror image of her Twitter feed, and both aim to give followers a warm-and-fuzzy feeling about the candidate. Clinton's predicament is a painful one. She's spent years filing the burrs off her personality, only to find that rough edges are in. The Hillary playlist, upbeat and bland, sounds anachronistic in a campaign that's more punk than pop.

News organizations, too, tend to be slow to adapt to the arrival of a new medium. Television, with its diurnal "news cycle," gave a theatrical rhythm to campaigns. Each day was an act in a broader drama that arced from conflict to crisis to resolution. Campaigns were "narratives." They had "story lines." Social media is different. Its fragmented messages and conversations offer little in the way of plot. Its literary style is stream of consciousness, more William Burroughs than William Thackeray. But reporters and pundits, stuck in the TV era, keep trying to fit the bits and pieces on Twitter and Facebook into a linear tale. That's why today's campaign reports often seem out of sync with the public's reaction to events.

Think of what happened in July when Trump went after John McCain. "He's not a war hero," Trump said in an Iowa speech. "I like people who weren't captured." In any prior campaign, such snide criticism of an American veteran who had been tortured as a prisoner of war would have constituted a major "gaffe." It would have immediately triggered a narrative of sin, penance, and redemption. In this familiar plot, a trope of modern campaigns, the candidate is first pilloried, then required to make a heartfelt apology, and finally, after the sincerity of the apology is carefully weighed, either granted absolution or ordered off the stage. At which point a new narrative would begin.

That's the way the news media played the Trump attack. In print and on TV, the putative gaffe received saturation coverage, with the aghast press dutifully reprimanding the wayward Donald. "Will Trump's Smear of McCain Doom His Candidacy?" asked a *Newsweek* headline. But the narrative, to the media's surprise, never advanced. Far from apologizing, Trump kept attacking. The tweets piled up, the public's attention buzzed to newer things, and the drama died in its first act. With social media, we seem to have entered a post-narrative world of campaigning. And that greatly circumscribes the power of traditional media in stage-managing races. Rather than narrating stories, anchors are reduced to reading tweets.

THE INTERNET, we've often been told, is a force for "democratization," and what we've seen so far with the coverage of the 2016 race seems to prove the point. It's worth asking, though, what kind of democracy is being promoted. Early digital enthusiasts assumed that the web, by freeing the masses from TV news producers and other media gatekeepers, would engender a deeper national conversation. We the people would take control of the discussion. We'd go online to read position papers, seek out diverse viewpoints, and engage in spirited policy debates. The body politic would get fit.

It was a pretty thought, but it reflected an idealized view both of human nature and of communication media. Even a decade ago, in the heady days of the blogosphere, there were signs that online media promoted a mob mentality. People skimmed headlines and posts, seeking information that reinforced their biases and shunning contrary perspectives. Information gathering was more tribalistic than pluralistic. As the authors of one study, published in *Perspectives on Politics* in 2010, concluded, "blog authors tend to link to their ideological kindred and blog readers gravitate to blogs that reinforce their existing viewpoints." The internet inspired participation, but the participants ended up in "cloistered cocoons of cognitive consonance."

That probably shouldn't have been a surprise. The net reinforced the polarizing effect that mass media, particularly talk radio and cable news, had been having for many years. What is a surprise is that social media is turning out to be more encompassing and controlling, more totalizing, than earlier media ever was. The social networks operated by companies like Facebook, Twitter, and Google don't just regulate the messages we receive. They regulate our responses. They shape, through the design of their apps and their information-filtering regimes, the forms of our discourse.

When we go on Facebook, we see a cascade of messages determined by the company's News Feed algorithm, and we're provided with a set of prescribed ways to react to each message. We can click a Like button; we can share the message with our friends; we can add a brief comment. With the messages we see on Twitter, we're given buttons for replying, retweeting, and liking, and any thought we express has to fit the service's tight text limits. Google News gives us a series of headlines, emphasizing the latest stories to have received a cluster of coverage, and it provides a row of buttons for sharing the headlines on other platforms. All social networks impose these kinds of formal constraints, both on what we see and on how we respond. The restrictions have little to do with the public interest. They reflect the commercial interests of the companies operating the networks, as well as the limits of software programming.

Because it simplifies and speeds up interpersonal communication, the formulaic quality of social media is well suited to the banter that takes place among friends. Clicking a heart symbol may be the perfect way to judge the worth of an Instagrammed selfie. But when applied to political speech, the same constraints can be pernicious. Political discourse rarely benefits from templates and routines. It becomes most valuable when it involves careful deliberation, an attention to detail, and subtle and open-ended critical thought—the kinds of things that social media tends to frustrate rather than promote.

Over the next year, as the presidential campaign careens toward its conclusion, all of us—the public, the press, the candidates—will get an education in how national elections work in the age of social media. We may discover that the gates maintained by our new gatekeepers are narrower than ever.

From Politico Magazine
2015

WHY ROBOTS WILL ALWAYS NEED US

"HUMAN BEINGS ARE ASHAMED to have been born instead of made," wrote the philosopher Günther Anders in 1956. Our shame has only deepened as our machines have grown more adept.

Every day we're reminded of the superiority of computers. Self-driving cars are immune to distractions and road rage. Automatic trains don't speed out of control. Factory robots don't goof off. Algorithms don't suffer the cognitive biases that can cloud the judgments of doctors, accountants, and lawyers. Computers work with a speed and precision that make us look like bumbling slackers.

It seems obvious: The best way to get rid of human error is to get rid of humans. But that assumption, however fashionable, is itself erroneous. Our desire to liberate ourselves from ourselves is founded on a fallacy. We exaggerate the abilities of computers even as we give our own talents short shrift.

It's easy to see why. We hear of every disaster involving human fallibility—the chemical plant that exploded because the technician forgot to open a valve, the plane that fell from the sky because the pilot mishandled the yoke—but what we don't hear about are all the times that people use their expertise to avoid accidents or defuse risks. Pilots, physicians, and other professionals routinely navigate unexpected dangers with great aplomb but little credit. Even in our daily routines, we perform feats of perception and skill that lie far beyond the capacity of the sharpest computers. Google is quick to tell us about how few accidents its autonomous cars are involved in, but it doesn't trumpet the many times the cars' backup drivers have had to take the wheel to steer the machines out of danger.

Computers are wonderful at following instructions, but they're lousy at improvisation. They resemble, in the words of computer scientist Hector Levesque, "idiot savants" who are "hopeless outside their area of expertise." Their talents end at the limits of their programming. Human skill is less circumscribed. Think of Captain Sully Sullenberger landing that Airbus A320 on the Hudson River after its engines were taken out by a flock of geese. Born of deep experience in the real world, such intuition lies beyond calculation. If computers had the ability to be amazed, they'd be amazed by us.

While our own flaws loom large in our thoughts, we view computers as infallible. Their scripted consistency presents an ideal of perfection far removed from our own clumsiness. What we forget is that our machines are built by our own hands. When we shift work to a machine, we don't eliminate human agency and its potential for error. We transfer that agency into the machine's workings, where it lies concealed until something goes awry. Computers break down. They have bugs. They get hacked. And when let loose in the world, they face situations that their programmers didn't prepare them for. They work perfectly until they don't.

Many disasters blamed on human error actually involve chains of events that are initiated or aggravated by technological failures. Consider the 2009 crash of Air France Flight 447 as it flew from Rio de Janeiro to Paris. While passing through a storm over the Atlantic, the plane's airspeed sensors iced over. Without the velocity data, the autopilot couldn't perform its calculations. It shut down, abruptly shifting control to the pilots. Taken by surprise in a stressful situation, the aviators made mistakes. The plane, with 228 passengers and crew, plunged into the ocean.

The crash was a tragic example of what scientists call the automation paradox. When a computer takes over a job, the workers are left with little to do. Their attention drifts. Their skills, lacking exercise, atrophy. Then, when the computer fails, the humans flounder. Software designed to eliminate human error ends up making human error more likely.

In 2013, the Federal Aviation Administration reported that an overreliance on automation has become a major factor in air disasters and urged airlines to give pilots more opportunities to fly manually. The best way to make flying even safer than it already is, the agency's research suggests, may be to transfer some responsibility away from computers and back to people. Where humans and machines work in concert, more automation is not always better.

That's a lesson that the Toyota Motor Company, a leader in factory automation, has learned the hard way. In recent years, the carmaker has had to recall millions of vehicles to fix defects, putting a dent in its profits and tarnishing its prized reputation for quality. It now believes that its manufacturing problems stem from a loss of human insight and talent. "We need to become more solid and get back to basics," a company executive told Bloomberg News, "to sharpen our manual skills and further develop them." Toyota is now replacing some of the robots in its Japanese plants with skilled workers. Craftsmen are returning to the line.

We're in this together, our computers and ourselves. Even if engineers create automated systems that can handle every possible contingency—far from a sure bet when it comes to complex jobs—it will be years before the systems are fully in place. In aviation, it would take decades to replace or retrofit the thousands of planes in operation, all of which were designed to have pilots in their cockpits. The same goes for roads and rails. Infrastructure doesn't change overnight.

Instead of seeing computers as our replacements, we would be wise to view them as our partners, with abilities complementary to our own. Software can play an invaluable role in handling routine chores and helping people avoid mistakes, but it can't match the versatility or common sense that human minds are capable of. What we risk sacrificing, if we rush to curtail our involvement in difficult work, are the flashes of insight and inspiration that set us apart from machines.

The world is less accident-prone than it's ever been, thanks to human ingenuity, technical advances, and thoughtful laws and regulations. Computers can help sustain that progress. Recent train crashes, including the fatal Amtrak derailment earlier this month, might have

been prevented had automated speed-control systems been in operation. Automotive software that senses when drivers are sleepy and
sounds alarms can prevent wrecks. Diagnostic algorithms can deliver
valuable information to doctors and challenge their preconceptions,
leading to better patient care.

The danger in dreaming of a perfectly automated society is that it
makes such modest improvements seem less pressing—and less worthy
of investment. Why bother taking small steps forward, if utopia lies
just around the bend?

From the New York Times
2015

LOST IN THE CLOUD

IN THE SPRING OF 1997, the Library of Congress opened an ambitious exhibit featuring several hundred of the most historically significant items in its collection. One of the more striking of the artifacts was the "rough draught" of the Declaration of Independence. Over Thomas Jefferson's original, neatly penned script ran edits by John Adams, Benjamin Franklin, and other founding fathers. Words were crossed out, inserted, and changed, the revisions providing a visual record of debate and compromise. A boon to historians, the four-page manuscript provides even the casual viewer with a keen sense of the drama of a nation being born.

Imagine if the Declaration were composed today. It would almost certainly be written on a computer screen rather than with ink and paper, and the edits would be made electronically, through email exchanges or a file shared on the internet. If we were lucky, a modern-day Jefferson would turn on track-changes and print copies of the document as it progressed. We'd at least know who wrote what, even if the generic computer type lacked the expressiveness of handwriting. More likely, the digital file would come to be erased or rendered unreadable by changes in technical standards. We'd have the words, but the document itself would have little resonance.

Historian Abby Smith Rumsey was one of the curators of the Library of Congress exhibit, and the experience informs *When We Are No More*, her wide-ranging rumination on cultural memory. "A physical connection between the present and past is wondrously forged through the medium of time-stained paper," she writes. But that "distinctive visceral connection" with history may be much diminished, if not lost, when our cultural heritage is stored in sterile databases rather than in actual objects. Rumsey's book poses a vital question: As more and more of what we know, make, and experience is recorded as vaporous bits in the cloud, what exactly will we leave behind for future generations?

We tend to think of memory as a purely mental phenomenon, something ethereal that goes on inside our minds. But scientists are discovering that our senses and even our emotions play important roles in recollection and remembrance. Memory seems to have emerged in animals as a way to navigate and make sense of the world, and the faculty remains tightly tied to the physical body and its material surroundings. Just taking a walk can help unlock memory's archives, studies have shown.

Rumsey draws a powerful analogy to underscore memory's materiality. The greatest memory system, she reminds us, is the universe itself. Nature embeds history in matter. When, in the nineteenth century, scientists realized they could read nature's memory by closely examining the Earth and stars, we gained a much deeper understanding of the cosmos and our place in it. Geologists discovered that the strata in exposed rock tell the story of the planet's development. Biologists found that fossilized plants and animals reveal secrets about the evolution of life. Astronomers realized that by looking through a telescope they could see not only across great distances but far back in time, gaining a glimpse of the origins of existence.

Through such discoveries, Rumsey argues, people both revealed and refined their "forensic imagination," a subtle and creative way of thinking highly attuned to deciphering meaning from matter. We deploy that same imagination in understanding and appreciating our history and culture. The upshot is that the technologies a society uses to record, store, and share information will play a crucial role in determining the richness, or sparseness, of its legacy. To put a new spin on Marshall McLuhan's famous dictum, the medium is the memory.

Whether through cave paintings or Facebook posts, we humans have always been eager to record our experiences. But we've been far less zealous about safeguarding those records for posterity. In choosing among media technologies through the ages, people have tended to trade durability for transmissibility. Intent on our immediate needs, we prefer those media that make communication easier and faster, rather than the ones that offer the greatest longevity. And so the lightweight scroll supplants the heavy clay tablet, the instantaneous email supplants the slow-moving letter. A cave painting may last for millennia, but a Facebook post will get you a lot more Likes a lot more quickly.

We're now in the midst of the most far-reaching shift in media ever, as we rush to replace all manner of physical recordings with digital alternatives. The benefits are many, but so are the losses. "Digital memory is ubiquitous yet unimaginably fragile," Rumsey reports, "limitless in scope yet inherently unstable." All media are subject to decay, of course. Clay cracks, paper crumbles. What's different now is that our cultural memory is embedded in a complex and ever-shifting system of technologies. Any change in the system can leave the record unreadable. A book can sit on a shelf for hundreds of years and retain its legibility. All that's required to decode it is a pair of eyes. A digital file is far more fussy. Dependent on computers for decoding, it can disappear or turn to gibberish whenever operating systems, software applications, or document standards are revised.

All of us have experienced the evanescence of the digital. Web pages change by the day, leaving little or no trace of their earlier versions. Hyperlinks dead-end in 404 error pages, with their irritating "Not Found" notices. Internet services and social media sites shut down, their data disappearing with them. And as for opening that file you saved on a floppy disk with an obsolete software program: Well, good luck. If we're not careful, Rumsey warns, "the history of the twenty-first century will be riddled with large-scale blanks and silences."

Rumsey is clear about the dangers of our "ephemeral digital landscape," but she isn't a doomsayer. She believes that we can protect our cultural legacy for our descendants, even if that legacy ends up mainly in the form of immaterial bits. But, she stresses, we'll first need to overcome our complacency and start taking the long-term protection of valuable data seriously. We'll need a reinvigorated system of libraries and archives, spanning the public, private, and nonprofit sectors, that are adept at digital preservation. We'll need thoughtful protocols for determining what data needs to be saved and what can be discarded. And we'll need to ensure that control over our cultural heritage doesn't end up in the hands of a small group of commercial enterprises that focus on profit, not posterity.

Those are prudent suggestions. But, even as we continue down the path to a virtualized future, we shouldn't lose sight of the enduring value of the material artifact. We should make sure that there's always a place in the world for the eloquent object, the thing itself.

From the Washington Post
2016

THE DAEDALUS MISSION

On what wings dare he aspire?

—*William Blake*

1.

In 2008, Samuel O. Poore, a plastic surgeon who teaches at the University of Wisconsin's medical school, published an article in the *Journal of Hand Surgery* titled "The Morphological Basis of the Arm-to-Wing Transition." Drawing on evolutionary and anatomical evidence, he laid out a workable method for using the techniques of modern reconstructive surgery, including bone fusing and skin and muscle grafting, to "fabricate human wings from human arms." Although the wings, in the doctor's estimation, would not be capable of generating the lift needed to get a person off the ground, they might nonetheless serve "as cosmetic features simulating, for example, the nonfunctional wings of flightless birds."

We have always envied birds their wings. From angels to superheroes, avian-human hybrids have been fixtures of myth, legend, and art. In the ninth century, the celebrated Andalusian inventor Abbas ibn Firnas fashioned a pair of wings out of wood and silk, attached them to his back, covered the rest of his body in feathers, and jumped from a promontory. He avoided the fate of his forebear Icarus, but "in alighting," a witness reported, "his back was very much hurt." Leonardo da Vinci sketched scores of plans for winged, human-powered flying machines called ornithopters. Batman's pinion-pointed cape looms over popular culture. *Birdman* won the best picture Oscar in 2015. "Red Bull gives you wings," promise the energy drink's ads.

Dr. Poole considered his paper a thought experiment, and he ended it with an admonition: "Humans should remain human, staying on the ground pondering and studying the intricacies of flight while letting birds be birds and angels be angels." Not everyone shared his caution. Advocates of radical human enhancement, or transhumanism, found inspiration in the article. One of them, writing on a popular transhumanist blog, suggested that it might soon be possible to craft working human wings by combining surgical techniques with synthetic muscles and genetic modifications. "Many humans have wished they could fly," the blogger wrote. "There's nothing morally wrong with granting that wish." The post garnered more than seven hundred comments. "I WANT WINGS!!!!!!!!!!!!!!!" went a typical one. "For as long as i can remember i have been longing to feel the wind in my feathers" went another.

2.

When Nora Ephron decided to call her 2006 essay collection *I Feel Bad about My Neck*, she all but guaranteed herself a bestseller. Prone to wattling and wrinkling, banding and bagging, the neck has long been a focal point of people's discontent with their bodies. But it's hardly the only body part that provokes disappointment and frustration. From miserly hair follicles to yellowed toenails, from forgetful brains to balky bowels, the body seems intent on reminding us of its flaws and insufficiencies. One thing that sets us apart from our animal kin is our ability to examine our bodies critically, as if they were things separate from ourselves. We may not think of ourselves as Cartesians anymore, but we remain dualists when it comes to distinguishing the self from its physical apparatus. And it's this ability to envision our bodies as instruments that allows us to imagine ways we might remodel or retrofit our anatomies to better reflect our desires and ideals. Our minds are always drafting new blueprints for our bodies.

We're quick to associate body modification with primitive cultures—the stereotypical savage with the bone in his nose—but that's a self-flattering fancy, a way to feel enlightened and civilized at the expense of others. When it comes to fiddling with the human body, we make even the most brutish of our ancestors look like amateurs. We go under the blade for nose jobs, tummy tucks, breast augmentations, hair transplants, face lifts, butt lifts, liposuctions, and myriad other cosmetic surgeries. We smooth our skin with dermabrasion brushes or chemical peels, conceal wrinkles with injections of botulinum toxin or hyaluronic filler. We brighten our smiles with whiteners and veneers, implants and orthodontia. We tattoo, pierce, and scarify our flesh. We swallow drugs and other potions to fine-tune our moods, sharpen our thinking, bulk up our musculature, control our fertility, and heighten our sexual prowess and pleasure. If to be transhuman is to use technology to change one's body from its natural state, for ornamental or functional purposes, then we are all already transhuman.

But our tinkering, however impressive, is only a prelude. The ability of human beings to alter and augment themselves is set to expand enormously in the decades ahead, thanks to a convergence of scientific and technical advances in such areas as robotics, bioelectronics, genetic engineering, and pharmacology. Up to now, body modifications have tended to be decorative or therapeutic. They've been used to improve or otherwise change people's looks or to repair damage from illnesses or wounds. Rarely have they offered people ways to transcend the body's natural limits. The future will be different. Progress in the field broadly known as biotechnology promises to make us stronger, smarter, and fitter, with sharper senses and more capable minds and bodies. Transhumanists have good reason to be excited. By the end of the twenty-first century what it means to be human is likely to be very different from what it means today.

War and medicine are the crucibles of human enhancement. They're where the need is pressing, the money plentiful. Military researchers, building on recent refinements in prosthetic arms and legs, are testing so-called Iron Man suits—artificial exoskeletons worn inside uniforms—that give soldiers greater strength, agility, and endurance. Wearing one current version, a G.I. can run a four-minute mile while carrying a full load of gear. Prototypes of more sophisticated bionic armor, which can sharpen vision, enhance situational awareness, and regulate body temperature along with boosting mobility and muscle, are in testing by the U.S. Special Operations Command. The merging of man and machine is well under way.

That goes for gray matter, too. In 2014, DARPA, the military's R&D arm, established a well-financed Biological Technologies Office to work on the frontiers of human enhancement. The new division's broad portfolio includes a raft of ambitious neuroengineering projects aimed at bolstering mental skill and accomplishment on and off the battlefield. In the works are brain implants that, in the agency's words, "facilitate the formation of new memories and retrieval of existing ones," neural interfaces that "reliably extract information from the nervous system . . . at a scale and rate necessary to control complex machines," and centimeter-sized neural "modems" that allow high-speed, standardized data transmissions between brains and computers.

While neuroscientists are still a long way from understanding consciousness and thought, they are, as the DARPA projects suggest, having success in reverse engineering many cognitive and sensory functions. Whenever knowledge of the brain expands, so too do the possibilities for designing tools to manipulate and augment mental processes. Cochlear implants, which translate sound waves into electrical signals and transmit them to the brain's auditory nerve, have already given tens of thousands of deaf people the ability to hear. In 2013, the Food and Drug Administration approved the first retinal implant. It gives sight to the blind by wiring a digital camera to the optic nerve. Scientists at Case Western Reserve University are developing a brain chip that monitors and adjusts levels of neurotransmitters, like dopamine, that regulate brain function. The researchers say the chip, which has been successfully tested in mice, works like a "home thermostat" for mental states.

Many such neural devices are in the early stages of development, and most are designed to aid the sick or disabled. But neuroengineering is progressing swiftly, and there's every reason to believe that implants and interfaces will come to be used by healthy people to gain new and exotic talents. "Advances in molecular biology, neuroscience and material science are almost certainly going to lead, in time, to implants that are smaller, smarter, more stable and more energy-efficient," brain scientists Gary Marcus and Christof Koch explained in a 2014 *Wall Street Journal* article. "When the technology has advanced enough, implants will graduate from being strictly repair-oriented to enhancing the performance of healthy or 'normal' people." We'll be able to use them to improve memory, focus, perception, and temperament, the scientists wrote, and, eventually, to speed the development of manual and mental skills by automating the assembly of neural circuitry.

These examples all point to a larger truth, one that lies at the heart of the transhumanist project. The human species is, in form and function, subject to biological constraints. It changes at the glacial pace of evolution. As soon as we augment the body with machinery and electronics, we accelerate the speed at which it can change. We shift the time scale of physiological adaptation from the natural, measured in millennia, to the technological, which plays out over decades, years, or mere months. Biology, when seen from a human perspective, is more about stasis than change. But when it comes to technology, nothing stands still. What's rudimentary today can be revolutionary tomorrow.

3.

The changes wrought by prosthetics, implants, and other hardware will play out in plain view. Blurring the line between tools and their users, they will turn people into what science fiction writers like to call cyborgs. More profound may be the microscopic changes accomplished by manipulating chemical reactions within and between cells. Advances in neurobiology have made possible a new generation of psychoactive drugs that will give individuals greater control over how their minds work. Exploiting the recent discovery that memories are malleable—they seem to change each time they're recalled—researchers are testing drugs that can, by blocking chemicals involved in memory formation, delete or rewrite troubling memories as they're being retrieved by the mind. A study by two Dutch psychologists, published in the journal *Biological Psychiatry* in 2015, shows that similar "amnesic" medications may be able to erase deep-seated phobias, such as a fear of spiders or strangers, by scrubbing certain memories of their emotional connotations.

Also coming out of pharmaceutical labs are drugs that accelerate learning and heighten intelligence by speeding up neuronal activity, tamping down extraneous brain signals, or stimulating new connections among nerve cells. As with brain implants, these so-called smart drugs, or neuroenhancers, are intended to be used medicinally—to help children with Down syndrome do better at school or to combat mental decay in the elderly—but they also hold promise for making able-minded people cleverer. For years now, drugs designed to treat attention and sleep disorders, like Adderall and Provigil, have been used by students and professionals to sharpen their mental focus and increase their productivity. As new drugs for cognitive enhancement become available, they too will see widespread "off-label" use. Eventually, well-vetted neuroenhancers that don't produce severe side effects will be cleared for general use. The economic and social advantages of enhanced intelligence, however narrowly defined, will override medical and moral qualms. Cosmetic neurology will join cosmetic surgery as a consumer choice.

Then there's genetic engineering. The much discussed gene-editing tool CRISPR, derived from bacterial immune systems, has in just the past few years transformed genomic research. Scientists can rewrite genetic code with far greater speed and precision, and at far lower cost, than was possible before. In simple terms, CRISPR pinpoints a target sequence of DNA on a gene, uses a bacterial enzyme to snip the sequence out, and then splices a new sequence in its place. The inserted genetic material doesn't have to come from the same species. Scientists can mix and match bits of DNA from different species, creating real-life chimeras.

With thousands of academic and corporate researchers, not to mention scores of amateur biohackers, experimenting with CRISPR, progress in genome editing has reached a "breakneck pace," according to Jennifer Doudna, a University of California biochemist who helped develop the tool. Combined with ever more comprehensive genomic maps, CRISPR promises to expand the bounds of gene therapy, giving doctors new ways to repair disease-causing mutations and anomalies in DNA, and may allow transplantable human organs to be grown in pigs and other animals. CRISPR also brings us a step closer to a time when genetic engineering will be practicable for a variety of human enhancements, at both the individual and the species level. Wings are not out of the question.

Transhumanists are technology enthusiasts, and technology enthusiasts are not the most trustworthy guides to the future. Their speculations have a tendency to spiral into sci-fi fantasies. Some of the most hyped biotechnologies will fail to materialize or will fall short of expectations. Others will take longer to pan out than projections suggest. As innovation researchers Paul Nightingale and Paul Martin point out in an article in the journal *Trends in Biotechnology*, the translation of scientific breakthroughs into practical technologies remains "more difficult, costly and time-consuming" than is often supposed. That's particularly true of medical procedures and pharmaceutical compounds, which often require years of testing and tweaking before they're ready for the market. Although CRISPR is already being used to reengineer goats, monkeys, and other mammals—Chinese researchers have created beagles with twice the normal muscle mass—scientists believe that, barring rogue experiments, clinical testing on people remains years away.

But even taking a skeptical view of biotechnology, discounting wishful forecasts of immortality, designer babies, and computer-generated superintelligence, it's clear that we humans are in for big changes. The best evidence is historical rather than hypothetical. In just the past decade, many areas of biotechnology, particularly those related to genomics and computing, have seen extraordinary gains. The advances aren't going to stop, and experience suggests they're more likely to accelerate than to slow. Whether techniques of radical human enhancement arrive in twenty years or fifty, they will arrive, and in their wake will come newer ones that we have yet to imagine.

In 1923, the English biologist J. B. S. Haldane gave a lecture before the Heretics Society in Cambridge on how science would shape humanity in the future. His view was optimistic, if warily so. He surveyed advances in physics that were likely "to render life more and more complex, artificial, and rich in possibilities." He suggested that chemists would soon discover psychoactive compounds that would "add to the amenity of life and promote the expression of man's higher faculties." But it was the biological sciences, he predicted, that would bring the greatest changes. Progress in understanding the functioning of the body, the working of the brain, and the mechanics of heredity was setting the stage for "man's gradual conquest" of his own physical and mental being. "We can already alter animal species to an enormous extent," he observed, "and it seems only a question of time before we shall be able to apply the same principles to our own."

Society would, Haldane felt sure, defer to the scientist and the technologist in defining the boundaries of the human species. "The scientific worker of the future," he concluded, "will more and more resemble the lonely figure of Daedalus as he becomes conscious of his ghastly mission, and proud of it."

4.

LeBron James is an illustrated man. The NBA star has covered his body with some forty tattoos, each selected to symbolize an aspect of his life or beliefs. The phrase "CHOSEN 1" is inscribed in a heavy gothic script across his upper back. On his chest, spanning his pectorals, is an image of a manticore, the legendary winged lion with the face of a man. Across his biceps runs the motto "What we do in life echoes in eternity," words spoken by Maximus, the Russell Crowe character, in the movie *Gladiator*. It wasn't long ago that tattoos were considered distasteful or even grotesque, the marks of drunken sailors, carnival geeks, and convicts. Now they're everywhere. Americans spend well over a billion dollars a year at tattoo parlors, more than a third of young adults sport at least one tattoo, and celebrities like James take pride in branding themselves with elaborate and evocative ink. The taboo has gone mainstream.

That's the usual trajectory for body modifications. First we recoil from them, then we get used to them, then we embrace them. In his talk, Haldane acknowledged that society will initially resist any new attempt to refashion human beings:

> There is no great invention, from fire to flying, which has not been hailed as an insult to some god. But if every physical and chemical invention is a blasphemy, every biological invention is a perversion. There is hardly one which, on first being brought to the notice of an observer from any nation which has not previously heard of their existence, would not appear to him as indecent and unnatural.

In time, attitudes change. Custom cures queasiness. What was seen as a perversion comes to be viewed as a remedy or a refinement, decent and even natural. The shapeshifter, once a pariah, becomes a pioneer, a hero.

Society's changing feelings toward tattoos is a fairly trivial example of such cultural adaptation. More telling is the way views of sex-reassignment procedures, the most drastic of commonly performed body modifications, have evolved. Voluntary sex-change operations date back at least to the ancient world, when they amounted to little more than clumsy castrations, but it wasn't until the 1930s that advances in surgical techniques and hormone therapies made sex reassignment technologically viable. Through the middle years of the twentieth century, when sex-change procedures remained rare, medically controversial, and fraught with legal obstacles, Americans tended to view transsexuals as at best freaks and at worst perverts. But the stigma dissipated through the latter years of the century, as sex-reassignment therapies became more sophisticated and routine and societal perceptions and norms changed. Today, although transsexuals still face prejudice, the public is coming to view sex-change procedures, whether surgical or chemical, not as treatments for unfortunate medical disorders but as ways to bring one's body into alignment with one's true identity. When Olympian Bruce Jenner came out as Caitlyn Jenner in the spring of 2015, she was feted by the media and praised by the president.

The history of transsexuality, writes Yale historian Joanne Meyerowitz in *How Sex Changed*, illuminates our times. Not only does it demonstrate "the growing authority of science and medicine," but it also "illustrates the rise of a new concept of the modern self that placed a heightened value on self-expression, self-improvement, and self-transformation." The perception of gender as a matter of inclination rather than biology, as a spectrum of possibilities rather than an innate binary divide, remains culturally and scientifically contentious. But its growing acceptance, particularly among the young, reveals how eager we are, whenever science grants us new powers over our bodies' appearance and workings, to redefine human nature as malleable, as a socially and personally defined construct rather than an expression of biological imperatives. Advances in biotechnology may be unsettling, but in the end we welcome them because they give us greater autonomy in remaking ourselves into what we think we *should* be.

Transhumanism is "an extension of humanism," argues Nick Bostrom, an Oxford philosophy professor who has been one of the foremost proponents of radical human enhancement. "Just as we use rational means to improve the human condition and the external world, we can also use such means to improve ourselves, the human organism. In doing so, we are not limited to traditional humanistic methods, such as education and cultural development. We can also use technological means that will eventually enable us to move beyond what some would think of as 'human.'" The ultimate benefit of transhumanism, in Bostrom's view, is that it expands "human potential," giving individuals greater freedom "to shape themselves and their lives according to their informed wishes." Transhumanism unchains us from our nature.

Other transhumanists take a subtly different tack in portraying their beliefs as part of the humanistic tradition. They suggest that the greatest benefit of radical enhancement is not that it allows us to transcend our deepest nature but rather to fulfill it. "Self-reconstruction" is "a distinctively human activity, something that helps define us," writes Duke University bioethicist Allen Buchanan in his book *Better Than Human*. "We repeatedly alter our environment to suit our needs and preferences. In doing this we inevitably alter ourselves as well. The new environments we create alter our social practices, our cultures, our biology, and even our identity." The only difference now, he says, "is that for the first time we can *deliberately*, and in a *scientifically informed way*, change our selves." We can extend the Enlightenment into our cells.

Critics of radical human enhancement, often referred to as bioconservatives, take the opposite view, arguing that transhumanism is antithetical to humanism. Altering human nature in a fundamental way, they contend, is more likely to demean or even destroy the human race than elevate it. Some of their counterarguments are pragmatic. By tinkering with life, they warn, researchers risk opening a Pandora's box, inadvertently unleashing a biological or environmental catastrophe. They also caution that access to expensive enhancement procedures and technologies is likely to be restricted to economic or political elites. Society may end up riven into two classes, with the merely normal masses under the bionic thumbs of an oligarchy of supermen. They worry, too, that as people gain prodigious intellectual and physical abilities, they'll lose interest in the very activities that bring pleasure and satisfaction to their lives. They'll suffer "self-alienation," as New Zealand philosopher Nicholas Agar puts it.

But at the heart of the case against transhumanism lies a romantic belief in the dignity of life as it has been given to us. There is an essence to humankind, bioconservatives believe, from which springs both our strengths and our defects. Whether bestowed by divine design or evolutionary drift, the human essence should be cherished and protected as a singular gift, they argue. "There is something appealing, even intoxicating, about a vision of human freedom unfettered by the given," writes Harvard professor Michael J. Sandel in *The Case against Perfection*. "But that vision of freedom is flawed. It threatens to banish our appreciation of life as a gift, and to leave us with nothing to affirm or behold outside our own will." As a counter to what they see as misguided utilitarian utopianism, bioconservatives counsel humility.

The transhumanists and the bioconservatives are wrestling with the largest of questions: Who are we? What is our destiny? But their debate is a sideshow. Intellectual wrangling over the meaning of humanism and the fate of humanity is not going to have much influence over how people respond when offered new means of self-expression and self-definition. The public is not going to approach transhumanism as a grand moral or political movement, a turning point in the history of the species, but rather as a set of distinct products and services, each offering its own possibilities. Whatever people sense is missing in themselves or their lives they will seek to acquire with whatever means available. And as standards of beauty, intelligence, talent, and status change, even those wary of human enhancement will find it hard to resist the general trend. Wherever they may lead us, our attempts to change human nature will be governed by human nature.

We are myth makers as well as tool makers. Biotechnology allows us to merge these two instincts, giving us the power to refashion the bodies we have and the lives we lead to more closely match those we imagine for ourselves. Transhumanism ends in a paradox. The rigorously logical work that scientists, doctors, engineers, and programmers are doing to enhance and extend our bodies and minds is unlikely to raise us onto a more rational plane. It promises, instead, to return us to a more mythical existence, as we deploy our new tools in an effort to bring our dream selves more fully into the world. LeBron James's body art, with its richly idiosyncratic melding of Christian and pagan iconography, all filtered through a pop-culture sensibility, feels like an augury.

"I want to fly!" cries Icarus in the labyrinth. "And so you shall," says Daedalus, his father, the inventor. It's an old story, but we're still in it, playing our parts.

5.

Just before twilight on a Saturday evening in the spring of 2015, the famed rock climber Dean Potter walked with his girlfriend, Jen Rapp, and his buddy, Graham Hunt, from a parking area along Glacier Point Road to the edge of Taft Point, some three thousand feet above the Merced River in Yosemite Valley. Potter and Hunt were planning a BASE jump. They would leap from a ledge near the point and glide in their wingsuits for a quarter mile over the valley before passing through a notch in a ridgeline near a rock outcropping called Lost Brother. They would then unfurl their parachutes and come in for a landing in a clearing on the valley floor. Rapp would serve as spotter and photographer.

BASE jumping, one of the more extreme of extreme sports, is banned in national parks. But Hunt and Potter were dedicated daredevils who didn't put much stock in rules. They had been jumping for years from cliffs and peaks throughout Yosemite, including the iconic Half Dome, and they had wingsuited from Taft Point several times, together and separately. The course they set for themselves that evening was dangerous—the notch was narrow, the winds contrary—but they were confident in their skills and their equipment. Potter held the mark for the longest wingsuit flight on record, having covered nearly five miles in a 2011 jump from the Eiger in Switzerland, and he had been featured in a National Geographic documentary called *The Man Who Could Fly*. Hunt, too, was considered one of the world's top jumpers.

The first wingsuiter, if you don't count Abbas ibn Firnas, was Franz Reichelt. A Vienna-born tailor who ran a dressmaking shop in Paris, he designed and stitched his own "parachute suit," as he called the winged garment. He tested it by jumping off the Eiffel Tower on February 4, 1912. The suit failed, and the fall killed him. More than eighty years passed before a Finnish company called BirdMan International began manufacturing reliable wingsuits and selling them to skydivers and BASE jumpers. Constructed of lightweight, densely woven nylon, modern wingsuits sheath the jumper's entire body, forming two wings between the arms and torso and another between the legs. By greatly expanding the surface area of the human frame, the suits create enough lift to allow a person to glide downward for several minutes while controlling trajectory through slight movements of the shoulders, hips, and knees. Wingsuiters frequently reach speeds of a hundred miles an hour or more, giving them an exhilarating sense that they're actually flying.

Potter and Hunt reached the launching spot near Taft Point around seven o'clock and zipped themselves into their wingsuits. Potter jumped first, followed quickly by Hunt, while Rapp shot pictures from a few yards away. The two jumpers dropped like stones for a couple of seconds before their suits filled with air. Then, their bodies buoyant, they soared across the mountain sky with wings outstretched, like a pair of giant, brightly colored birds. "Part of me says it's kind of crazy to think you can fly your human body," Potter had told a *New York Times* reporter a few years earlier. "Another part of me thinks all of us have had the dream that we can fly. Why not chase after it? Maybe it brings you to some other tangent."

Jen Rapp kept taking photographs until Potter and Hunt passed through the notch and out of sight. She thought she heard something, a couple of thumps, but she told herself it was probably just the chutes opening. She waited for the text message that would let her know the pair had landed safely. Nothing came. Her phone was silent. Park rangers recovered the bodies the next morning.

2016

ACKNOWLEDGMENTS

THE SEVENTY-NINE POSTS COLLECTED in the first section of this book appeared between 2005 and 2015 in my blog Rough Type. Many of them have been reworked—compressed, expanded, or otherwise tinkered with—to fit the book's form and flow. A few of the posts also appeared, in different versions, in other publications: "Looking into a See-Through World" appeared in *The Guardian* (under the title "The Internet Rewards the Lazy and Punishes the Intrepid"); "Nowness" appeared in *Nieman Reports* (under the title "News in the Age of Now"); "The Medium Is McLuhan" appeared in *The New Republic*; "The Hierarchy of Innovation" and "Will Gutenberg Laugh Last?" appeared in the *Wall Street Journal* (under the respective titles "Why Our Innovators Traffic in Trifles" and "Don't Burn Your Books—Print Is Here to Stay"); "Out of Control" appeared in the book *What to Think about Machines That Think* (under the title "The Control Crisis"); "Our Algorithms, Ourselves" appeared in the *Los Angeles Review of Books* (under the title "The Manipulators"); and "The Seconds Are Just Packed" appeared in the book *What* Should *We Be Worried About?* (under the title "The Patience Deficit").

The aphorisms in the second section of the book were also drawn from Rough Type.

The third section includes works originally published, sometimes in different forms, in the following newspapers, magazines, and books: "Flame and Filament" was the epilogue to my book *The Big Switch: Rewiring the World, from Edison to Google*; "Is Google Making Us Stupid?" and "Mother Google" (under the title "Googlethink") appeared in *The Atlantic*; "Screaming for Quiet" (under the title "Deafened"), "Hooked," and "Past-Tense Pop" appeared in *The New Republic*; "The

Dreams of Readers" appeared in the U.K. collection *Stop What You're Doing and Read This!*, published by Vintage; "Life, Liberty, and the Pursuit of Privacy" (under the title "Tracking Is an Assault on Liberty, with Real Dangers") appeared in the *Wall Street Journal*; "The Library of Utopia" appeared in *MIT Technology Review*; "The Boys of Mountain View" (under the title "Growing Up Google") appeared in *The National Interest*; "The Eunuch's Children" (under the title "Paper Versus Pixel") appeared in *Nautilus*; "The Love That Lays the Swale in Rows" was the concluding chapter of my book *The Glass Cage: How Our Computers Are Changing Us*; "Why Robots Will Always Need Us" appeared in the *New York Times*; "The Snapchat Candidate" (under the title "How Social Media Is Ruining Politics") appeared in *Politico Magazine*; and "Lost in the Cloud" (under the title "When Our Culture's Past Is Lost in the Cloud") appeared in the *Washington Post*. "The Daedalus Mission" was written for this volume.

I am grateful to the many talented editors, proofreaders, and fact checkers that I have had the pleasure to work with, among them Don Peck, Isaac Chotiner, Sheila Glaser, Ryan Sager, Rachel Dry, Brian Bergstein, Charles Arthur, Justine Rosenthal, Michele Pridmore-Brown, Luba Ostashevsky, James Gibney, Natalie Shutler, Patrick Appel, Frances MacMillan, Randy Rothenberg, Gary Rosen, James Bennet, Lisa Kalis, Warren Bass, Jason Pontin, Parul Sehgal, Barbara Kiser, Lucas Wittman, John McMurtrie, Greg Veis, Nick Thompson, Luke Mitchell, Steve Levingston, Juliet Lapidos, Melissa Ludtke, Megan Garber, Mark Armstrong, Amy Bernstein, Art Kleiner, Reihan Salam, Tim Lavin, Mike Orcutt, and Sue Parilla.

I am fortunate to have Brendan Curry, at W. W. Norton, as my editor and John Brockman, at Brockman Inc., as my literary agent. And let me thank my wife, Ann, for her patience in putting up with my blogging, even when I was tapping out posts in the small hours of the morning. I deserved worse.

INDEX

Abbas ibn Firnas, 329, 341
Abedin, Huma, 315
Abercrombie & Fitch, 244–45
accessibility, 99–100, 199–200, 268
    instantaneous, 57, 232, 241, 264, 267
    of music, 293
Adams, John, 325
Adderall, 304
Addiction by Design (Schüll), 218–19
Adorno, Theodor, 153–54
advertising, 15, 31, 168, 255, 258, 264
    edginess in, 10–11
    as pervasive, 64
    search-linked, 279–80
    in social media, 53–54
    in virtual world, 27
    see also marketing
Advisory Council on the Right to Be
        Forgotten, 194
AdWords, 279
aesthetic emotions, 249–50
Against Intellectual Monopoly (Levine), 276
Agar, Nicholas, 339
Agarwal, Anant, 133
air disasters, 322–23
Air France Flight 447, 322
Akamai Technologies, 205
"Alastor" (Shelley), 88
Alfred P. Sloan Foundation, 272
algorithms, 113, 136, 145, 167, 174, 190–94,
        237, 238, 242, 257, 258
allusion, cultural nuances of, 86–89
alphabet, ideograms vs., 234
Altamont concert, 42
AltaVista, 67
amateurs, 33
    creativity of, 49
    internet hegemony of, 4–8
    media production by, 81
Amazon, 31, 37–38, 92, 142, 256, 277, 288
ambient overload, 90–92
America Online, 279–80

"Amorality of Web 2.0, The" (Carr), xxi–xii
Amtrak derailment, 323
analog resources, 148–50
Anders, Günther, 321
Anderson, Chris, 68
Andreessen, Marc, xvii
Andrews, James, 134
Android phones, 156, 283
anticonsumerism, 83–85
anxiety, 186, 304
Apple, 125
Apple Corps, 71
Apple II, 76–77
archiving, cultural memory and, 325–28
Arendt, Hannah, 310–11
Aristotle, 174, 307–9
art:
    allusion in, 89
    bundling of musical tracks as, 42–43
    by-number, 71–72
    digitalization of, 223
    emotional response to, 249–50
    "free" vs. "servile," 253
    games, 261
    industrialization of, 209
    Instagram and, 224
    technology vs., 300
    transformative mirrors and, 131
Art as Experience (Dewey), 304
artificial intelligence (AI), 55, 145, 224, 231,
        261
    dangers of, 187–89
    neural networks for, 136–37
    vs. human intelligence, 69, 187, 239–40,
        242
artificial language, 215
Ask Jeeves, 280
Association of American Publishers, 141, 269
astronomy, 326
Atlantic, 8, 49, 249
attention deficit hyperactivity disorder, 304
attention disorders, video games and, 95–96

attention economy, cash economy vs., 31
attentiveness, 93–97
  diffusion of, 231–32, 236–37
  loss of, 322
audiobooks, 142, 287
Authors Guild, 269
autism, 102
automation:
  anesthetizing effect of, 305–7
  benefits of, 321
  economic effects of, 174–77, 310
  human interaction in, 321–24
  in parenting, 181
  utopian promise of, 308–11
  vehicular, 195–98
automation paradox, 322
automobiles, 226
  and noise pollution, 243–44
  self-driving, 195–98, 321, 324
avatars, 26, 39–40, 114
  conflict among, 25, 260
AW (after the web), 60–61

Barber, Elinor, 12–13
Barger, Jorn, 21
Barlow, John Perry, 85
Barrow Gurney, England, 56
BASE jumping, 341–42
"battle of speeds," 44
Baudrillard, Jean, 35–36
Bavelier, Daphne, 93–94
Beatles, The: Rock Band, 71–72
behavioralism, 211–13
Being Digital (Negroponte), xx
Bell, Daniel, 235
Beniger, James, 188
Benkler, Yochai, xviii
Bennett, William, 133
Berkman Center for Internet and Society, 271–73
"beta sprint," 272
Better Than Human (Buchanan), 339
Bezos, Jeff, 172
Big Brother, 105–6
big data, 150, 163, 211–13
Bilton, Nick, 94, 152–54
bioconservatives, 339–40
Biological Technologies Office, 332
bionic armor, 332
biotechnology, 331–40
BioWare, 261
BirdMan International, 341
Bishop, Elizabeth, 160
Bissell, Tom, 260–63
Blake, William, 329

Bloch, Bradley, 81
Bloglines, 67
blogs, blogging, 13–15, 35, 76–77, 79, 186,
    232, 252, 273, 319, 330
  author's experience in, xv–xvi, xxi–xii, 7–8,
    66–67
  as term, 21–22
Blow, Jonathan, 261
Blue Velvet (film), 34
Bodkin, Tom, 237
body modification, 330–31
Boesel, Whitney Erin, 157
Boing Boing, 14–15, 76
Bonforte, Jeff, 195
books, 91, 248
  concentration and, 232–34
  continuing sales of, 288, 291
  copying of, 122–25
  e-readers vs., 74, 110–11, 122–23, 140–43,
    225, 252–54, 257, 287–91
  newspapers vs., 79
  predicted obsolescence of, 79, 252, 287–88
  psychological effects of, 250–51
  repositories of, see libraries
  scanning devices for, 268–69
  spell of, 247–54
  as transformed by printing press, 103,
    240–41, 286
bootlegging (music), 121–25
boredom, 137
  abundance and, 217–18
Botsman, Rachel, 83–84
Bowker, 141, 142
Boyle, Richard, 12–13
Boy's Will, A (Frost), 297, 305
Braid (video game), 261
brain:
  artificial, 23
  bioengineering of, 332–34
  computer interface with, 239–40
  computerized monitoring of, 149–50
  in construction of knowledge, 199–200
  effect of virtual world on, 303–4
  internet-induced functional changes in,
    231–42
  as malleable, 235–36
  of McLuhan, 102, 106
  in reading, 247–54, 289–90
  in thought-shared messaging, 214–15
branding, internet opportunities for, 53, 108
Brin, Sergey, 23, 131, 239, 279, 284–85
  personal style of, 16–17, 281–82
Britannica Online, 5, 8
Brown, Jerry, 171

Buchanan, Allen, 339
Buddhist meditation, 162
bundling, of music, 41–46
Burroughs, William, 318
Bush, Jeb, 317
BW (before the web), 60–61
Byrne, David, 136

"C30 C60 C90 Go!," 121
Cai Lun, 286–87, 291
candlelight, 229–30
capitalism, 83–85
Case against Perfection, The (Sandel), 340
Case of Emily V., The (Oatley), 248
cassette tape, 121, 124
CCTV cameras, 52
CDs (compact disks), 42, 123–24, 293
cell phones, 52, 80, 233
censorship:
    in China, 283
    free flow of information vs., 191
Centers for Disease Control and Prevention,
    304
centrifugal force, 67
centripetal force, 66
Chambers, John, 134
Chen, Steve, 29
Chief Officers of State Library Agencies, 272
Chin, Denny, 269, 272
China, censored searches in, 283
Christian, Rebecca, 80
citation, allusion vs., 87–88
Clash, 63–64
classical music, 43–44
Claude Glass, 131–32
Clinton, Bill, 315
Clinton, Hillary, 314, 315, 317–18
clocks, changes wrought by, 235–36
clones, virtual, 26–27
cloud computing programs, 264, 283
cloud storage, 163, 168, 185, 225
    physical archives vs., 326
CNET, 55
Coachella festival, 126
Coca-Cola, marketing of, 53–54
cocaine, 262
cochlear implants, 332
cognitive bias, 321
cognitive control, 96
cognitive function:
    effect of internet on, 199–200, 231–42
    effect of video games on, 93–97
    "flow" state in, 297
    memory and, 98–99

neuroengineering of, 332
reading and, 248–52
cognitive surplus, 59–60
    avoidance of, 74
Coleridge, Samuel Taylor, 251
Collaborative Consumption, 84–85, 148
Columbia Records, 43–44
commercialism:
    anticonsumerism and, 83–85
    culture transformed by, xvii–xxii, 3, 9, 150,
        177, 198, 214–15
    in innovation, 172
    of libraries, 270–71
    media as tool of, 106, 213, 240, 244–45,
        257–58, 320
    in virtuality, 25–27, 72
commodes, high-tech, 23–24
communication:
    between computers, 167
    computer vs. human, 152–54
    evolution of, 53
    loneliness and, 159
    mass, 67–68
    speed of, 223, 320
    thought-sharing in, 214–15
Communist Manifesto (Marx and Engles),
    308
"Complete Control," 63–64
Computer Power and Human Reason
    (Weizenbaum), 236
computers:
    author's early involvement with, xix–xi
    benefits and limitations of, 322–23
    in education, 134
    effect on paper consumption of, 287–88
    evolution of, xix–x, 165
    future gothic scenarios for, 112–15
    human hybridization with, 37–38, 332
    human partnership with, 321–24
    as impediment to knowledge perception,
        303–4
    minds uploaded to, 69
    revivification through, 69–70
    written word vs., 325–28
concentration, diffusion of, 231–33, 236–37
Confession d'un Enfant du Siècle, La (Musset),
    xxiii
Congress, U.S., 275–77
consumer choice, 44–45
Consumer Electronics Show (CES), 32, 56
consumerism:
    counterculture co-opted by, 72
    distraction and, 65
    media as tool of, 106, 132, 219

consumption, self-realization vs., 64–65
contemplation, 241, 246
    through work, 298–99
conversation, computer streaming of, 152–54
CopyBot controversy, 25–27
copyright laws:
    history of, 275–76
    in online library controversies, 269–71,
        275–78, 283
    in virtual world, 25–27
Corporate Communalists, 83
corporate control, through self-tracking, 163–65
correspondence courses, 133–34
cosmetic surgery, 331, 334
Costeja González, Mario, 190–92, 194
Coupland, Douglas, 102, 103
Courant, Paul, 270, 272
courtesy:
    decline of, 157
    inefficiency of, 152–54
Cowen, Tyler, 116
Crawford, Matthew, 265
creativity, 49, 64
    before the virtual world, 60–61
    economics of, 8–9
    in music, 44–45, 294
    stifled by iPad, 76–78
    see also innovation
"crisis of control," 188–89
CRISPR, 334–35
crowdsourcing, 37
Cruz, Ted, 314
cultural memory, archiving of, 325–28
cutouts (remaindered record albums), 122
CyberLover, 55
cybernetics, 37–38, 214
cyberpunk, 113
cyberspace, xvii, 127
    early idealism of, 85
"Cyborg Manifesto" (Haraway), 168–69
cyborgs, 131
cynicism, 158

Daedalus, 336, 340
Darnton, Robert, 270–75, 278
DARPA, 332
Dash Express, 56
data-mining, 186, 212, 255–59
data-protection agencies, 190–91
Data Protection Directive, 191, 193
Davidson, Cathy, 94
Davies, Alex, 195
Davies, William, 214–15
Dean, Jeff, 137

death, as hardware failure, 115
Declaration of Independence, 278, 325
"Declaration of the Independence of
        Cyberspace" (Barlow), 85
deep reading, 241
deletionists, 18–20, 58
democratization, xvi, xviii, 28, 86, 89, 115,
        208, 271
    internet perceived as tool for, 319–20
depression, 304
Derry, N.H., 296–97
Descartes, René, 301, 330
Dewey, John, 304
"digital dualism," 129
"digital lifestyle," 32–33
digital memory, 327
digital preservation, 325–28
Digital Public Library of America (DPLA),
        268, 271–78
"Digital Republic of Letters," 271
discovery, adventure of, 13–15
Disenchanted Night (Schivelbusch), 229
displaced agency, 265
distraction, xix, 14, 316
    in consumerism, 65
    video games and, 19
diversity, 65
DNA, 69–70, 334–35
Doctorow, Cory, 76–77
"Does the 'New Economy' Measure Up to the
        Great Inventions of the Past?" (Gordon),
        116–17
Doors, 126
dopamine, 332
dot-com crash, xvi
Doudna, Jennifer, 335
driving, 195–98
Droit-Volet, Sylvie, 203–4
drones, 306
Drucker, Peter, 182
drugs, 119, 304, 331
    psychoactive, 333–34
    video games and, 262
    virtual, 39–40
Drum, Kevin, 306
Dylan, Bob, 121, 294
dystopias, 108

ears, development and evolution of, 235
Earthlink, 280
"Easter, 1916" (Yeats), 88
Easton, David, 211
"E-book Reading Jumps; Print Book Reading
        Declines," 140

ebooks, e-reader devices, 74, 122–23, 140–43, 225, 257, 274, 288, 290
   reading experience transformed by, 252–54
economic gap, xix, 30–31, 176–77, 179
economy, effect of technology on, 174–77, 179–80
Edison, Thomas, xvii, 134, 229, 287
education, technological transformation of, 133–35
Edwards, Douglas, 280–82, 285
efficiency:
   in computer communication, 152–54
   maximizing of, 84–85, 148, 164–65, 195–97, 209, 214, 234, 237–39, 303, 305
   of robots, 321
Eiffel Tower, 341
e-learning fad, 134
election campaigns:
   of 2008, 314
   of 2016, 314–20
   transformed by technology, 314–20
Electric Kool-Aid Acid Test, The (Wolfe), 170
Eliot, T. S., 86–87, 144
Ellison, Larry, 17
Eloi, 114, 186
Elster, Jon, 64–65
email, 34, 73, 91, 134, 186
emancipation:
   central control vs., 165
   computer technology perceived as path to, xvii–xix, 3, 11, 310
   tools as, 308
   see also liberation mythology
embodied cognition, 297
Emerson, Ralph Waldo, 98, 100, 247–48, 254, 313
emojis, 167, 215
emoticons, 30, 215
emotional intelligence (EI), 162
EmoTree, 168
empathy, 251–52
Encyclopedia Britannica, 110–11
encyclopedias:
   open-source, 5–8, 110
   traditional, 8, 14, 19, 110–11
   see also Wikipedia
endless-ladder myth, 174–77
"End of Books, The" (Uzanne), 287–88
"End of the Future, The" (Thiel), 116
England's Dreaming (Savage), 63
Enlightenment, Age of, 271–72, 339
Ephron, Nora, 331
ethics, 48, 226

technology and, 304–11, 329–42
Etzler, John Adolphus, xvi–xvii
Europeana, 272
European Union, 280
Everything Bad Is Good for You (Johnson), 13, 93–94
Everything Is Miscellaneous (Weinberger), 41
Exile on Main Street (album), 42–43, 45
"Exposure" (Heaney), xxii
Extra Lives (Bissell), 260–63
eyeglasses, reality augmented by, 108–9, 131–32, 160–61

Facebook, xv, xvi, 30–31, 50, 106, 113, 115, 119, 137, 138, 155–59, 166, 178, 186, 197, 205, 210, 223, 257, 265, 269, 284
   cynicism of, 158
   marketing through, 53–54
   political use of, 314, 317–20
   privacy and, 107, 193
   as record, 326–27
Facebook Home, 156–59
Facebook Social Advertising Event, 53
fact-mongering, 58–62
factory production, 308
   efficiency in, 164–65, 237–38, 305
fads, 71–72
Faithfull, Marianne, 42
fallibility, human vs. computer, 321–23
farming, 296–98
   technological advancement in, 305–6
Farrell, Thomas, 186
Faster (Gleick), 204
Favela Chic, 113–14
Federal Aviation Administration, 322–23
Federal Trade Commission, U.S. (FTC), 280, 284
feedback loops, 67
Feldman, Morton, 216
Ferriero, David, 272
fiction, effect on brain of, 248–52
filters, information overload and, 90–92
Finnegans Wake (Joyce), 106
first nature, 179–80
Fitbit, 119, 197
Flickr, xvi
flight, human quest for, 329–30, 340–42
Fonda, Jane, Wikipedia entry on, 6–7
Food and Drug Administration, 332
Foreman, Richard, 241–42
forensic imagination, 326
45 rpm singles, 44–46, 121
Four-Second Rule, 205
Foursquare, 257

Fox, Justin, 116
France, Google and, 264
Franklin, Benjamin, 325
Friedman, Bruce, 232–33
Friedman, Thomas, 133
Frost, Robert, 145–46, 182, 247–48, 296–99,
    302, 304–5, 313
Fuller, Buckminster, 171

Galbraith, John Kenneth, xix
Gates, Bill, xvii
    Jobs compared to, 32–33
    Wikipedia entry on, 5–6
Gelernter, David, xvii
gender reassignment, 337–38
generational change, 230
generativity, 76–78
gene therapy, 335
genetic engineering, 334–35
geology, 326
Germany, Google and, 283–84
GIFs, 203
Gil, Sandrine, 203–4
Gilligan's Island, fact-mongering about,
    58–62
Gillmor, Dan, 7
Gleick, James, 204
Go, GeForce (avatar), 25
Goldsmith, Kenneth, 216–17
Google, 13, 67, 79, 86–89, 112, 115, 144–46,
    162, 181, 195, 199, 204, 205, 226, 253,
    257, 321
    in AI, 136–37
    competition for, 284–85
    corporate management of, 16–17
    customized searching on, 264–66
    early days of, xvii, 279–81
    effect on memory of, 98–101
    ethical criticism of, 283
    failed projects of, 269, 283
    goals of, 23–24, 87, 145–46, 239–40, 268
    growth and evolving hegemony of, 279–
        85
    international projects of, 283–84
    investigations into, 280, 284–85
    music streaming by, 207, 209
    in online privacy case, 190–94
    philosophy of, 279–80, 283
    political use of, 319
    social stream management by, 166–67
    universal book project of, 267–72, 275–77,
        283
Google Apps, 283
Google Blog Search, 66

Google Book Search, 268–72, 275–77, 283
Google bus, 170–71, 173
"Google Effects on Memory" (Sparrow, Liu,
    and Wegner), 98
Google Glass, 131–32, 160–61, 164
Google Maps, 153
Google News, 315, 320
Google Now, 145
Google Play Music, 207
Googleplex, 17, 238
    restrooms of, 23–24
Google Reader, 67
Google Serendipity, 13, 15
Google Suggest, 264–65
Google X lab, 195
Gordon, Robert J., 116–17
Gothic High-Tech, 113–15
GPS systems, 56–57, 226, 304
Graham, Lindsay, 314
Grand Theft Auto, 262
Gray, John, 36
Great Man theory, 28–29
Green, Shawn, 93–94
Greenfield, Patricia, 95
Grimmelmann, James, 277
Grossman, Lev, 28–29
Grover, Monte, 185
Guitar Hero (game), 64–65
Gutenberg Galaxy, The (McLuhan), 102–3

hackability, 76–78
HAL (computer), 231, 239, 242
Haldane, J. B. S., 336
Haraway, Donna, 168–69
hardware failure, 112–15
Harmonix Music Systems, 71
Harper, William Rainey, 133–34
Hart, Michael, 278
Harvard University, library of, 269, 270
HathiTrust, 277
Hawking, Stephen, 210
health:
    bioengineering of, 331–33
    computer monitoring of, 149–50, 164,
        324
Heaney, Seamus, xxii, 201
hearing loss, 245–46
Hedi, Emperor of China, 286
Heim, Michael, xvii
Hellman, Monte, 203
Hendrix, Janie, 126
Hendrix, Jimi, 126
Hennessy, John, 133
Here Comes Everybody (Shirky), 61

Herrera, Luis, 273
hierarchy of innovation, 117–20
hierarchy of needs, 117–19
higher-value tasks, 174–75, 177
Hillis, Danny, 23
Hilton, Paris, 10
history, documentation of, 325–27
Hitachi Business Microscope, 163–64
Hof, Robert, 157
Holland, Norman, 251, 254
Hollerith, Herman, 188
holograms, 126, 109
Homer, 292
Hopkins, Gerard Manley, xxi
hormone therapy, 337–38
Horning, Rob, 64–65
"How Google Is Changing Your Brain" (Ward and Wegner), 200
How Sex Changed (Meyerowitz), 338
Hoyt, Clark, 47–48
Huber, Jeff, 195
Huffington Post, 81, 186
human beings:
    as "analog resources," 148–50
    aspiration as basic to, 340
    biological constraints of, 333
    codification of, 37–38
    eclipsed by technology, 108–9, 187–89, 239–42, 308–9
    essence of, 310–13, 339–40
    mind vs. body in, 23–24
    radical enhancement of, 330–40
    superior qualities of, 321–24
Human Condition, The (Arendt), 310–11
humanism, transhumanism and, 338–39
human potential movement, 4
Hunt, Graham, 341–42
Hurley, Chad, 29
Huth, John Edward, 301
hyperlinks, 67, 87, 88, 231–42, 327
hypertext, 14, 223

IBM, 26
Icarus, 329, 340
ideograms, alphabet vs., 234
I Feel Bad About My Neck (Ephron), 330
I Live in the Future & Here's How It Works (Bilton), 94
image recognition, 137
I'm Feeling Lucky (Edwards), 281, 285
immediacy, 57, 70, 234
immortality, 69–70, 115, 210
    see also resurrection
impatience, technology-induced, 203–6

implants, 332–33
improvisation, 322
inclusionists, 18–20
industrial revolution, 77, 174–75, 237–38
"informal experience," 197–98
information overload, 241
    filters in, 90–92
information storage:
    biological vs. digital, 200
    cloud-based, 163, 168, 185, 225
information technology, dangers of, 187–89, 190–94
Innis, Harold, 103
"innocent fraud," xix
innovation:
    as American tradition, 171–72
    at Google, 281–82
    hierarchy of, 118–20
    nostalgia vs., 292
    obstacles to, 278
    shift in focus of, 116–20
In Pursuit of Silence (Prochnik), 243–44
Instagram, 166, 186, 224, 314, 320
instant gratification, 206
Instant Messaging, 34
intellectual technologies, 235–36
intelligence, effect of internet on, 231–42
interactivity, 106, 223
    of e-readers, 252–53
interface, 216–19
internal clocks, 203–4
internet:
    beneficial aspects of, 231–32
    biases reinforced by, 319–20
    centralization of, 66–68
    commercial aspects of, xvi–xxi, 3, 9, 83–85, 150, 240, 257–58, 320
    control of, xx
    criminal use of, 55, 257–58
    in education, 134
    effect on paper consumption of, 286
    evolution of, 3–4, 225
    as free, 8–9
    human beings reprogrammed by, 237
    idealistic prediction for, 3–4, 9
    in illusion of knowledge, 199–200
    intellectual technologies subsumed into, 236–37
    liberation mythology of, 41–42
    manipulation of memory on, 47–48
    personal data collected and monitored on, see data-mining
    political uses of, 314–20
    regulation of, 190–94

internet (*continued*)
  as restrictive vs. expansive, 8
  technical glitches of, 66–67
  traffic analysis of, 30
  see also Web 2.0, Web 1.0; *specific platforms*
Internet Archive, 272, 277
Introduction to Karl Marx, An (Elster), 64
intuition, 322
inventions, 116–17, 229–30, 287, 301, 305–6
iPad, 74, 142, 289
  closed nature of, 76–78
iPhone, 113, 149
  children's apps on, 74
  closed nature of, 76
  introduction of, 32–33
iPod, 33, 125, 197, 217, 245, 287
Ireland, Google and, 284
"IRL Fetish, The" (Jurgenson), 127
Iron Man suits, 331
Isaacson, Walter, 121
isolation, paradox of connection and, 35–36,
    159, 184, 255
iTunes, 41, 42, 125

Jacobs, Alan, 14
Jagger, Mick, 42, 292
James, LeBron, 336–37, 340
James, Rick, 126
James, William, 203
Jampol, Jeff, 126
Jarvis, Jeff, 252
Jefferson, Thomas, xvii, 271, 306, 325
Jenner, Caitlyn, 338
Jennings, Leslie, 16
Jensen, Brennen, 72
Jobs, Steve, 32–33, 76, 113, 115, 121, 162
Johnson, Steven, 13–15, 83–84, 93–94
Jones, Brian, 42
Jones, Mick, 63
Joplin, Janis, 126
Joyce, James, 106
Jurgenson, Nathan, 127–29
Justice Court, European Union, online pri-
    vacy case in, 191–92
Justice Department, U.S., 269

Kahle, Brewster, 272, 277
Karp, Scott, 10–11, 232
Katriel, Tamar, 186
Keller, Michael, 272
Kelly, Kevin, 4, 5, 8–9
Kennedy, John, 315, 317
Kesey, Ken, 170–71, 173
Keynes, John Maynard, 306, 310

Kidd, David Comer, 252
Kindle, 122, 142–43, 257, 277, 288
Kindle Fire, 142
Kirsch, Adam, 86–87, 89
Kittler, Friedrich A., 235
"Kitty Hawk" (Frost), 299
Knight Capital, 187–88
knowledge:
  desire vs., 313
  illusion of, 199–200, 224
  for its own sake, 253
  through action, 297–304, 313
  wisdom vs., 240
knowledge work, 176, 238
Koch, Christof, 333
Korzybski, Alfred, 303
Kostelanetz, Richard, 184
Kraus, Allen, 47
Kubrick, Stanley, 108, 231, 242
Kurzweil, Fredric, 69–70
Kurzweil, Ray, 49, 69–70, 145

"Lady with the Dog, The" (Chekhov), 250
Lamartine, Alphonse de, 79
language, natural, displaced by digital, 201–2,
    214–15
Larkin, Philip, 159, 186
Latour, Bruno, 179–80
Lawrence, D. H., xxiii, 50
Leary, Timothy, 171, 172
Leaves of Grass (Whitman), 20
Left 4 Dead, 260
Lehrer, Jonah, 94, 95
leisure, technologies of, 119–20
Leonard, Sarah, 178
Levchin, Max, 148–50
Levesque, Hector, 322
Levine, David K., 276
Levy, Steven, 3–4
Lewis, Wyndham, 103
liberation mythology, 41, 59–61, 308–11
libraries:
  archiving in, 325, 327
  budgetary challenges to, 273
  online, 267–78
Library of Congress, 325–26
"library of utopia," 267
libsecondlife, 26
Licklider, J. C. R., 306–7
lightbulbs, 183
lighting, advancements in, 229–30
Lim, Kevin, 26
Lindbergh, Charles, 306
Linden, Robin, 26

Linden Lab, 26
LinkedIn, xvi, 166, 186
Listen.com, 122
"Literary Allusion in the Age of Google"
    (Kirsch), 86
literature, allusion and, 86–89
Literature and the Brain (Holland), 251
"Little Gidding" (Eliot), 144
Liu, Jenny, 98
long-playing (LP) record albums, 41–44,
    121
Lord, Albert, 103
Lorraine, Claude, 131
Losse, Kate, 178
Louth, Andrew, 253
love, 225
    unified theory of, 210, 213
Lowrey, Annie, 174
Ludd, Ned, 77, 178
Luddites, 76–78, 106, 202, 241, 312
Lynch, David, 108

Macfarlane, Robert, 201–2
machine intelligence, 136–37
    language of, 214–15
Machine in the Garden, The (L. Marx), 131
machine zone, addiction to, 218–19
Macrowikinomics (Tapscott and Williams),
    84
magazines, online vs. printed, 288–89, 291
Mailer, Norman, 102
mainstream media, blogging vs., 7–8
Malick, Terrence, 155
"Man-Computer Symbiosis" (Licklider),
    306–7
Manjoo, Farhad, 195
Mann, Horace, 12
Man of the Year, "you" as, 28–29
Man Who Could Fly, The (film), 341
maps, digital, 56–57
Mar, Raymond, 250
Marcus, Gary, 333
marketing:
    for Facebook Home, 156–59
    through social media, 53–54
    on YouTube, 108–9
Marr, David, 212
*Marshall McLuhan: You Know Nothing of My
    Work!* (Coupland), 102
Martin, Paul, 335
Marx, Karl, Marxism, xvii, xviii, 26, 83, 174,
    308
Marx, Leo, 131
Maslow, Abraham, 117–20

massive open online courses (MOOCs), 133
master-slave metaphor, 307–9
mastery, 64–65
Mayer, Marissa, 268
Mayer-Schönberger, Viktor, 48
McAfee, Andrew, 195
McCain, John, 318
McKeen, William, 13–15
McLuhan, Marshall, 102–6, 183–84, 232,
    326
McNealy, Scott, 257
measurement, 182
    of experience, 197–98, 211–12
mechanical loom, 77
Mechanical Turk, 37–38
media:
    as advertorial, 53
    big outlets for, 67
    changes in, 53–54, 59–60
    democratization of, xvi, xviii, 28
    hegemony of internet in, 236–37
    intellectual and social effects of, 103–6
    as invasive, 105–6, 127–30
    mainstream, 7–8
    pursuit of immediacy in, 79
    real world vs., 223
    in shaping thought, 232
    smartphones' dominance of, 183–84
    tools vs., 226
meditation, 162
Mehta, Mayank, 303
memory:
    association and cohesion in, 100–101
    computer, 147, 231
    cultural, 325–28
    digital, 327
    effect of computers on, 98–101, 234, 240
    internet manipulation of, 48
    neuroengineering of, 332–34
    packaging of, 186
    in revivification, 69–70
    spatial, 290
    time vs., 226
    video games and, 94–97
Merholz, Peter, 21
Merleau-Ponty, Maurice, 300
Merton, Robert, 12–13
message-automation service, 167
Meyer, Stephenie, 50
Meyerowitz, Joanne, 338
microfilm, microphotography, 267
Microsoft, 108, 168, 205, 284
military technology, 331–32
Miller, Perry, xvii

mindfulness, 162
Minima Moralia (Adorno), 153–54
mirrors, 138–39
Mitchell, Joni, 128
Mollie (video poker player), 218–19
monitoring:
  corporate control through, 163–65
  of thoughts, 214–15
  through wearable behavior-modification
    devices, 168–69
Montaigne, Michel de, 247, 249, 252, 254
Moore, Geoffrey, 209
Morlocks, 114, 186
"Morphological Basis of the Arm-to-Wing
  Transition, The" (Poore), 329–30
Morrison, Ewan, 288
Morrison, Jim, 126
Morse code, 34
"Most of It, The" (Frost), 145–46
motor skills, video games and, 93–94
"Mowing" (Frost), 296–300, 302, 304–5
MP3 players, 122, 123, 124, 216, 218, 293
multitasking, media, 96–97
Mumford, Lewis, 138–39, 235
Murdoch, Rupert and Wendi, 131
music:
  bundling of, 41–46
  commercial use of, 244–45
  copying and sharing technologies for,
    121–26, 314
  digital revolution in, 293–95
  fidelity of, 124
  listening vs. interface in, 216–18, 293
  in participatory games, 71–72
  streamed and curated, 207, 217–18
music piracy, 121–26
Musings on Human Metamorphoses (Leary),
  171
Musk, Elon, 172
Musset, Alfred de, xxiii
Muzak, 208, 244
MySpace, xvi, 10–11, 30–31

"Names of the Hare, The," 201
nanotechnology, 69
Napster, 122, 123
narcissism, 138–39
  Twitter and, 34–36
narrative emotions, 250
natural-language processing, 215
Negroponte, Nicholas, xx
neobehavioralism, 212–13
Netflix, 92
neural networks, 136–37

neuroengineering, 332–33
New Critics, 249
News Feed, 320
news media, 318–20
newspapers:
  evolution of, 79, 237
  online archives of, 47–48, 190–92
  online vs. printed, 289
Newton, Isaac, 66
New York Public Library, 269
New York Times, 8, 71, 83, 133, 152–53, 195,
    237, 283, 314, 342
  erroneous information revived by, 47–48
  on Twitter, 35
Nielsen Company, 80–81
Nietzsche, Friedrich, 126, 234–35, 237
Nightingale, Paul, 335
Nixon, Richard, 317
noise pollution, 243–46
Nook, 257
North of Boston (Frost), 297
nostalgia, 202, 204, 312
  in music, 292–95
Now You See It (Davidson), 94

Oates, Warren, 203
Oatley, Keith, 248–50
Obama, Barack, 314
obsession, 218–19
OCLC, 276
"off grid," 52
Olds, James, 235
O'Neill, Gerard, 171
One Infinite Loop, 76
Ong, Walter, 129
online aggregation, 192
On Photography (Sontag), xx
open networks, profiteering from, 83–85
open-source projects, 5–7, 26
Oracle, 17
orchises, 305
O'Reilly, Tim, 3–5, 7
organ donation and transplantation, 115
ornithopters, 239
orphan books, 276, 277
Overture, 279–80
Owad, Tom, 256
Oxford Junior Dictionary, 201–2
Oxford University, library of, 269

Page, Larry, 23, 160, 172, 239, 268–69, 270,
    279, 281–85
  personal style of, 16–17, 281–82, 285
paint-by-number kits, 71–72

Paley, William, 43
Palfrey, John, 272–74, 277
Palmisano, Sam, 26
"pancake people," 242
paper, invention and uses of, 286–89
Paper: An Elegy (Sansom), 287
Papert, Seymour, 134
Paradise within the Reach of All Men, The
        (Etzler), xvi–xvii
paradox of time, 203–4
parenting:
    automation of, 181
    of virtual child, 73–75
Parker, Sarah Jessica, 131
participation:
    "cognitive surplus" in, 59
    as content and performance, 184
    inclusionists vs. deletionists in, 18–20
    internet, 28–29
    isolation and, 35–36, 184
    limits and flaws of, 5–7, 62
Paul, Rand, 314
Pendragon, Caliandras (avatar), 25
Pentland, Alex, 212–13
perception, spiritual awakening of, 300–301
personalization, 11
    of ads, 168, 225, 264
    isolation and, 29
    loss of autonomy in, 264–66
    manipulation through, 258–59
    in message automation, 167
    in searches, 145–46, 264–66
    of streamed music, 207–9, 245
    tailoring in, 92, 224
    as threat to privacy, 255
Phenomenology of Perception (Merleau-Ponty),
        300
Philosophical Investigations (Wittgenstein), 215
phonograph, phonograph records, 41–46,
        133, 287
photography, technological advancement in,
        311–12
Pichai, Sundar, 181
Pilgrims, 172
Pinterest, 119, 186
playlists, 314
PlayStation, 260
"poetic faith," 251
poetry, 296–313
polarization, 7
politics, transformed by technology, 314–20
Politics (Aristotle), 307–8
Poore, Samuel O., 329–30
pop culture, fact-mongering in, 58–62

pop music, 44–45, 63–64, 224
    copying technologies for, 121–26
    dead idols of, 126
    industrialization of, 208–9
    as retrospective and revivalist, 292–95
positivism, 211
Potter, Dean, 341–42
power looms, 178
Presley, Elvis, 11, 126
Prim Revolution, 26
Principles of Psychology (James), 203
Principles of Scientific Management, The
        (Taylor), 238
printing press:
    consequences of, 102–3, 234, 240–41,
        271
    development of, 53, 286–87
privacy:
    devaluation of, 258
    from electronic surveillance, 52
    family cohesion vs., 229
    free flow of information vs. right to, 190–
        94
    internet threat to, 184, 255–59, 265, 285
    safeguarding of, 258–59, 283
    vanity vs., 107
proactive cognitive control, 96
Prochnik, George, 243–46
"Productivity Future Vision (2011)," 108–9
Project Gutenberg, 278
prosperity, technologies of, 118, 119–20
prosumerism, 64
protest movements, 61
Proust and the Squid (Wolf), 234
proximal clues, 303
public-domain books, 277–78
"public library," debate over use of term,
        272–74
punch-card tabulator, 188
punk music, 63–64

Quantified Self Global Conference, 163
Quantified Self (QS) movement, 163–65
Quarter-of-a-Second Rule, 205

racecars, 195, 196
radio:
    in education, 134
    evolution of, 77, 79, 159, 288
    as music medium, 45, 121–22, 207
    political use of, 315–16, 317, 319
Radosh, Daniel, 71
Rapp, Jen, 341–42
reactive cognitive control, 96

Readers' Guide to Periodical Literature, 91
reading:
 brain function in, 247–54, 289–90
 and invention of paper, 286–87
 monitoring of, 257
 video gaming vs., 261–62
 see also books
reading skills, changes in, 232–34, 240–41
Read Write Web (blog), 30
Reagan, Ronald, 315
real world:
 digital media intrusion in, 127–30
 perceived as boring and ugly, 157–58
 as source of knowledge, 313
 virtual world vs., xx–xxi, 36, 62, 127–30,
  303–4
reconstructive surgery, 239
record albums:
 copying of, 121–22
 jackets for, 122, 224
 technology of, 41–46
Redding, Otis, 126
Red Light Center, 39
Reichelt, Franz, 341
Reid, Rob, 122–25
relativists, 20
religion:
 internet perceived as, 3–4, 238
 for McLuhan, 105
 technology viewed as, xvi–xvii
Republic of Letters, 271
reputations, tarnishing of, 47–48, 190–94
Resident Evil, 260–61
resource sharing, 148–49
resurrection, 69–70, 126
retinal implants, 332
Retromania (Reynolds), 217, 292–95
Reuters, Adam, 26
Reuters' SL bureau, 26
revivification machine, 69–70
Reynolds, Simon, 217–18, 292–95
Rice, Isaac, 244
Rice, Julia Barnett, 243–44
Richards, Keith, 42
"right to be forgotten" lawsuit, 190–94
Ritalin, 304
robots:
 control of, 303
 creepy quality of, 108
 human beings compared to, 242
 human beings replaced by, 112, 174, 176,
  195, 197, 306–7, 310
 limitations of, 323
 predictions about, xvii, 177, 331

replaced by humans, 323
 threat from, 226, 309
Rogers, Roo, 83–84
Rolling Stones, 42–43
Roosevelt, Franklin, 315
Rosen, Nick, 52
Rubio, Marco, 314
Rumsey, Abby Smith, 325–27
Ryan, Amy, 273

Sandel, Michael J., 340
Sanders, Bernie, 314, 316
Sansom, Ian, 287
Savage, Jon, 63
scatology, 147
Schachter, Joshua, 195
Schivelbusch, Wolfgang, 229
Schmidt, Eric, 13, 16, 238, 239, 257, 284
Schneier, Bruce, 258–59
Schüll, Natasha Dow, 218
science fiction, 106, 115, 116, 150, 309, 335
scientific management, 164–65, 237–38
Scrapbook in American Life, The, 185
scrapbooks, social media compared to, 185–86
"Scrapbooks as Cultural Texts" (Katriel and
  Farrell), 186
scythes, 302, 304–6
search-engine-optimization (SEO), 47–48
search engines:
 allusions sought through, 86
 blogging, 66–67
 in centralization of internet, 66–69
 changing use of, 284
 customizing by, 264–66
 erroneous or outdated stories revived by,
  47–48, 190–94
 in filtering, 91
 placement of results by, 47–48, 68
 searching vs., 144–46
 targeting information through, 13–14
 writing tailored to, 89
 see also Google
searching, ontological connotations of,
  144–46
Seasteading Institute, 172
Second Life, 25–27
second nature, 179
self, technologies of the, 118, 119–20
self-actualization, 120, 340
 monitoring and quantification of, 163–65
selfies, 224
self-knowledge, 297–99
self-reconstruction, 339
self-tracking, 163–65

Selinger, Evan, 153
serendipity, internet as engine of, 12–15
SETI@Home, 149
sexbots, 55
Sex Pistols, 63
sex-reassignment procedures, 337–38
sexuality, 10–11
    virtual, 39
Shakur, Tupac, 126
sharecropping, as metaphor for social media,
    30–31
Shelley, Percy Bysshe, 88
Shirky, Clay, 59–61, 90, 241
Shop Class as Soulcraft (Crawford), 265
Shuster, Brian, 39
sickles, 302
silence, 246
Silicon Valley:
    American culture transformed by, xv–xxii,
        148, 155–59, 171–73, 181, 241, 257, 309
    commercial interests of, 162, 172, 214–15
    informality eschewed by, 197–98, 215
    wealthy lifestyle of, 16–17, 195
Simonite, Tom, 136–37
simulation, *see* virtual world
Singer, Peter, 267
Singularity, Singularitarians, 69, 147
sitcoms, 59
situational overload, 90–92
skimming, 233
"Slaves to the Smartphone," 308–9
Slee, Tom, 61, 84
SLExchange, 26
slot machines, 218–19
smart bra, 168–69
smartphones, xix, 82, 136, 145, 150, 158,
    168, 170, 183–84, 219, 274, 283, 287,
    308–9, 315
Smith, Adam, 175, 177
Smith, William, 204
Snapchat, 166, 205, 225, 316
social activism, 61–62
social media, 224
    biases reinforced by, 319–20
    as deceptively reflective, 138–39
    documenting one's children on, 74–75
    economic value of content on, 20–21,
        53–54, 132
    emotionalism of, 316–17
    evolution of, xvi
    language altered by, 215
    loom as metaphor for, 178
    maintaining one's microcelebrity on, 166–
        67
    paradox of, 35–36, 159
    personal information collected and moni-
        tored through, 257
    politics transformed by, 314–20
    scrapbooks compared to, 185–86
    self-validation through, 36, 73
    traditional media slow to adapt to, 316–19
    as ubiquitous, 205
    see also specific sites
social organization, technologies of, 118, 119
Social Physics (Pentland), 213
Society for the Suppression of Unnecessary
    Noise, 243–44
sociology, technology and, 210–13
Socrates, 240
software:
    autonomous, 187–89
    smart, 112–13
solitude, media intrusion on, 127–30, 253
Songza, 207
Sontag, Susan, xx
SoundCloud, 217
sound-management devices, 245
soundscapes, 244–45
space travel, 115, 172
spam, 92
Sparrow, Betsy, 98
Special Operations Command, U.S., 332
speech recognition, 137
spermatic, as term applied to reading, 247,
    248, 250, 254
Spinoza, Baruch, 300–301
Spotify, 293, 314
"Sprite Sips" (app), 54
Squarciafico, Hieronimo, 240–41
Srinivasan, Balaji, 172
Stanford Encyclopedia of Philosophy, 68
Starr, Karla, 217–18
Star Trek, 26, 32, 313
Stengel, Rick, 28
Stephenson, Neal, 116
Sterling, Bruce, 113
Stevens, Wallace, 158
Street View, 137, 283
Stroop test, 98–99
Strummer, Joe, 63–64
Studies in Classic American Literature
    (Lawrence), xxiii
Such Stuff as Dreams (Oatley), 248–49
suicide rate, 304
Sullenberger, Sully, 322
Sullivan, Andrew, xvi
Sun Microsystems, 257
"surf cams," 56–57

surfing, internet, 14–15
surveillance, 52, 163–65, 188–89
surveillance-personalization loop, 157
survival, technologies of, 118, 119
Swing, Edward, 95

Talking Heads, 136
talk radio, 319
Tan, Chade-Meng, 162
Tapscott, Don, 84
tattoos, 336–37, 340
Taylor, Frederick Winslow, 164, 237–38
Taylorism, 164, 238
Tebbel, John, 275
Technics and Civilization (Mumford), 138, 235
technology:
    agricultural, 305–6
    American culture transformed by, xv–xxii,
        148, 155–59, 174–77, 214–15, 229–30,
        296–313, 329–42
    apparatus vs. artifact in, 216–19
    brain function affected by, 231–42
    duality of, 240–41
    election campaigns transformed by, 314–20
    ethical hazards of, 304–11
    evanescence and obsolescence of, 327
    human aspiration and, 329–42
    human beings eclipsed by, 108–9
    language of, 201–2, 214–15
    limits of, 341–42
    master-slave metaphor for, 307–9
    military, 331–32
    need for critical thinking about, 311–13
    opt-in society run by, 172–73
    progress in, 77–78, 188–89, 229–30
    risks of, 341–42
    sociology and, 210–13
    time perception affected by, 203–6
    as tool of knowledge and perception, 299–
        304
    as transcendent, 179–80
Technorati, 66
telegrams, 79
telegraph, Twitter compared to, 34
telephones, 103–4, 159, 288
television:
    age of, 60–62, 79, 93, 233
    and attention disorders, 95
    in education, 134
    Facebook ads on, 155–56
    introduction of, 103–4, 159, 288
    news coverage on, 318
    paying for, 224
    political use of, 315–16, 317

technological adaptation of, 237
    viewing habits for, 80–81
Teller, Astro, 195
textbooks, 290
texting, 34, 73, 75, 154, 186, 196, 205, 233
Thackeray, William, 318
"theory of mind," 251–52
Thiel, Peter, 116–17, 172, 310
"Things That Connect Us, The" (ad cam-
    paign), 155–58
30 Days of Night (film), 50
Thompson, Clive, 232
thought-sharing, 214–15
"Three Princes of Serendip, The," 12
Thurston, Baratunde, 153–54
time:
    memory vs., 226
    perception of, 203–6
Time, covers of, 28
Time Machine, The (Wells), 114
tools:
    blurred line between users and, 333
    ethical choice and, 305
    gaining knowledge and perception through,
        299–304
    hand vs. computer, 306
    Home and Away blurred by, 159
    human agency removed from, 77
    innovation in, 118
    media vs., 226
    slave metaphor for, 307–8
    symbiosis with, 101
Tosh, Peter, 126
Toyota Motor Company, 323
Toyota Prius, 16–17
train disasters, 323–24
transhumanism, 330–40
    critics of, 339–40
transparency, downside of, 56–57
transsexuals, 337–38
Travels and Adventures of Serendipity, The
    (Merton and Barber), 12–13
Trends in Biochemistry (Nightingale and
    Martin), 335
TripAdvisor, 31
trolls, 315
Trump, Donald, 314–18
"Tuft of Flowers, A" (Frost), 305
tugboats, noise restrictions on, 243–44
Tumblr, 166, 185, 186
Turing, Alan, 236
Turing Test, 55, 137
Twain, Mark, 243
tweets, tweeting, 75, 131, 315, 319

language of, 34–36
  theses in form of, 223–26
"tweetstorm," xvii
20/20, 16
Twilight Saga, The (Meyer), 50
Twitter, 34–36, 64, 91, 119, 166, 186, 197,
  205, 223, 224, 257, 284
  political use of, 315, 317–20
2001: A Space Odyssey (film), 231, 242
Two-Lane Blacktop (film), 203
"Two Tramps in Mud Time" (Frost), 247–48
typewriters, writing skills and, 234–35, 237

Uber, 148
Ubisoft, 261
Understanding Media (McLuhan), 102–3, 106
underwearables, 168–69
unemployment:
  job displacement in, 164–65, 174, 310
  in traditional media, 8
universal online library, 267–78
  legal, commercial, and political obstacles
    to, 268–71, 274–78
universe, as memory, 326
Urban Dictionary, 145
utopia, predictions of, xvii–xviii, xx, 4, 108–9,
  172–73
Uzanne, Octave, 286–87, 290

Vaidhyanathan, Siva, 277
vampires, internet giants compared to,
  50–51
Vampires (game), 50
Vanguardia, La, 190–91
Van Kekerix, Marvin, 134
vice, virtual, 39–40
video games, 223, 245, 303
  as addictive, 260–61
  cognitive effects of, 93–97
  crafting of, 261–62
  violent, 260–62
videos, viewing of, 80–81
virtual child, tips for raising a, 73–75
virtual world, xviii
  commercial aspects of, 26–27
  conflict enacted in, 25–27
  language of, 201–2
  "playlaborers" of, 113–14
  psychological and physical health affected
    by, 304
  real world vs., xx–xxi, 36, 62, 127–30
  as restrictive, 303–4
  vice in, 39–40
von Furstenberg, Diane, 131

Wales, Jimmy, 192
Wallerstein, Edward, 43–44
Wall Street, automation of, 187–88
Wall Street Journal, 8, 16, 86, 122, 163,
  333
Walpole, Horace, 12
Walters, Barbara, 16
Ward, Adrian, 200
Warhol, Andy, 72
Warren, Earl, 255, 257
"Waste Land, The" (Eliot), 86, 87
Watson (IBM computer), 147
Wealth of Networks, The (Benkler), xviii
"We Are the Web" (Kelly), xxi, 4, 8–9
Web 1.0, 3, 5, 9
Web 2.0, xvi, xvii, xxi, 33, 58
  amorality of, 3–9, 10
  culturally transformative power of, 28–29
  Twitter and, 34–35
"web log," 21
Wegner, Daniel, 98, 200
Weinberger, David, 41–45, 277
Weizenbaum, Joseph, 236
Wells, H. G., 114, 267, 278, 288
What's Mine Is Yours (Botsman and Rogers),
  83–84
"When Every Song Ever Recorded Fits on
    Your MP3 Player, Will You Listen to
    Any of Them?" (Starr), 218
When We Are No More (Rumsey), 325–27
Whitman, Walt, 20, 183, 184
wicks, 229–30
Wiener, Anthony, 315
wiki, as term, 19
"wikinomics," 84
Wikipedia, xvi, 21, 192
  in fact-mongering, 58
  hegemony of, 68
  ideological split in, 18–20
  slipshod quality of, 5–8
wiki-sects, 18
Wilde, Oscar, 174, 308
Williams, Anthony, 84
Wilson, Fred, 11
Windows Home Server, 32
Winer, Dave, 35
wings, human fascination with, 329–30, 335,
  340–42
wingsuits, 341–42
Wired, xvii, xxi, 3, 4, 106, 156, 162, 174, 195,
  232
Wittgenstein, Ludwig, 215
Wolf, Gary, 163
Wolf, Maryanne, 234

Wolfe, Tom, 170
work:
    as basis for society, 310–11, 313
    in contemplative state, 298–99
    efficiency in, 165–66, 237–38
    job displacement in, 164–65, 174, 310
    trivial alternatives to, 64
World Brain (Wells), 267
World Health Organization, 244
World of Warcraft, 59
Wozniac, Steve "Woz," 32
Wright brothers, 299
writing:
    archiving of, 325–27
    and invention of paper, 286–87
writing skills, changes in, 231–32, 234–35,
    240

Xbox, 64, 93, 260
X-Ray Spex, 63

Yahoo, 67, 279–80
Yahoo People Search, 256
Y Combinator Startup School, 172
Yeats, William Butler, 88
Yelp, 31
Yosemite Valley, 341–42
youth culture, 10–11
    as apolitical, 294–95
    music and, 125
    TV viewing in, 80–81
YouTube, 29, 31, 58, 75, 81, 102, 186, 205,
    225, 314
    technology marketing on, 108–9

Zittrain, Jonathan, 76–77
zombies, 260, 263
Zuckerberg, Mark, xvii, xxii, 53, 115, 155,
    158, 215, 225
    Facebook Q & A session of, 210–11, 213, 214
    imagined as jackal, xv